International Conference on Industry 4.0 for Agri-food Supply Chains: Addressing Socio-economic and Environmental Challenges in Ukraine

International Conference on Industry 4.0 for Agri-food Supply Chains: Addressing Socio-economic and Environmental Challenges in Ukraine

Editors

Hana Trollman
Iuliia Samoilyk

MDPI

Basel • Beijing • Wuhan • Barcelona • Belgrade • Novi Sad • Cluj • Manchester

Editors
Hana Trollman
School of Business, University of Leicester
Leicester, UK

Iuliia Samoilyk
Economy and International Economic Relations Department, Poltava State Agrarian University
Poltava, Ukraine

Editorial Office
MDPI
St. Alban-Anlage 66
4052 Basel, Switzerland

This is a reprint of articles from the Proceedings published online in the open access journal *Engineering Proceedings* (ISSN 2673-4591) (available at: https://www.mdpi.com/2673-4591/40/1).

For citation purposes, cite each article independently as indicated on the article page online and as indicated below:

Lastname, A.A.; Lastname, B.B. Article Title. *Journal Name* **Year**, *Volume Number*, Page Range.

ISBN 978-3-0365-9388-3 (Hbk)
ISBN 978-3-0365-9389-0 (PDF)
doi.org/10.3390/books978-3-0365-9389-0

Cover image courtesy of Hana Trollman

Contents

About the Editors

Hana Trollman

Hana Trollman, Ph.D., is a Lecturer in Management at the University of Leicester School of Business, UK (since 2022). She studied sustainable manufacturing at the Wolfson School of Mechanical, Electrical, and Manufacturing Engineering at Loughborough University (Loughborough, UK), where she received her PhD (2021). Her research centers on the sustainability of food systems, particularly focusing on the supporting role of Industry 4.0 technologies. She is the principal investigator on the Universities UK (Research England) funded project "Traceability in food supply chains for improved global and national competitiveness" under the UK–Ukraine R&I twinning grants scheme (2023). She was awarded the Loughborough University Doctoral College Student Prize (2021) and is also a Fellow of the Higher Education Academy (FHEA) (2020).

Iuliia Samoilyk

Iuliia Samoilyk, Doctor of Economic Sciences, is a Full Professor at Poltava State Agrarian University, Poltava City, Ukraine. She teaches at the Department of Economics and International Economic Relations. Dr. Samoilyk has taught classes on international statistics, international agribusiness, enterprise strategy, strategy of the agrarian industry and enterprise development, and leadership.

Dr. Samoilyk holds a Master of Management from the Poltava State Agrarian Academy (currently, University) with a major in Organizational Management. She also has a Ph.D. in Economic Sciences from Zhytomyr National Agroecology University, Ukraine, and a post-doctorate in Economic Sciences from Poltava State Agrarian University.

She is an expert of the National Agency for Quality Assurance in Higher Education and a scholar of the "Hero of the Heavenly Hundred – the Hero of Ukraine, Igor Serdyuk" Heavenly Hundred Award. In 2021-2023, Dr. Samoilyk participated in a Collaborative Online International Learning (COIL) project with Dr. Zlatinka Blaber from Salem State University, Massachusetts, USA. She has also participated in a faculty exchange program at the University of Missouri, USA, funded by the USDA (2018). She has participated in a pilot project for the sustainable internationalization of Ukrainian research in the context of the globalization of the Ukrainian food sector (UaFoodTrade) by the Leibniz Institute of Agricultural Development in Transitional Economies and the Kyiv School of Economics, Germany–Ukraine (2019). Between 2022 and 2023, Dr. Samoilyk participated in the Academic Mobility Erasmus project at Polytechnic University in Cartagena, Spain. In 2023, she was co-investigator in an International project with the University of Leicester (UK), "Traceability in food supply chains for improved global and national competitiveness." She also participated in the "IAMO Forum – 2023" and the International IAMO DAAD Alumni Training for Research Managers (2023).

engineering proceedings

MDPI

Editorial

Preface of the International Conference on Industry 4.0 for Agri-food Supply Chains: Addressing Socio-economic and Environmental Challenges in Ukraine (IC4AFSC 2023) †

Hana Trollman [1,*] and Iuliia Samoilyk [2]

1 School of Business, University of Leicester, Leicester LE2 1RQ, UK
2 Economy and International Economic Relations Department, Poltava State Agrarian University, 36003 Poltava, Ukraine; iuliia.samoilyk@gmail.com
* Correspondence: ht203@leicester.ac.uk

† Presented at the International Conference on Industry 4.0 for Agri-food Supply Chains: Addressing Socio-economic and Environmental Challenges in Ukraine, Leicester, UK and Online, 24–25 July 2023.

Citation: Trollman, H.; Samoilyk, I. Preface of the International Conference on Industry 4.0 for Agri-food Supply Chains: Addressing Socio-economic and Environmental Challenges in Ukraine (IC4AFSC 2023). *Eng. Proc.* **2023**, *40*, 24. https://doi.org/10.3390/engproc2023040024

Published: 2 August 2023

The first edition of the International Conference on Industry 4.0 for Agri-food Supply Chains: Addressing Socio-economic and Environmental Challenges in Ukraine (IC4AFSC 2023) was held in hybrid form in Leicester, United Kingdom, from 24 to 25 July 2023.

The IC4AFSC 2023 conference has, as its underlying motivation, the aiding of Ukraine in the war against Russia by specifically addressing how Industry 4.0 technologies could benefit agri-food supply chains in terms of alleviating the socio-economic and environmental challenges precipitated by Russian aggression. The interdisciplinary conference serves as a forum for the exchange of novel ideas and presentation of the latest advancements and technological applications related to Industry 4.0 in agri-food supply chains from engineering (electric and electronic engineering, mechanical engineering, manufacturing engineering, and environmental engineering), business, and management perspectives, welcoming the global academic community and relevant industry into the discussion on current progress.

The scientific program of the conference consisted of six distinguished keynotes providing insights into recent developments and future prospects:

- Professor Louise Manning, University of Lincoln, UK, "Being resilient in challenging times in food supply chains";
- Professor Samir Dani, Keele University, UK, "Developing an agenda for Interdisciplinary Research within food risks and sustainability";
- Professor Wayne Martindale, University of Lincoln, UK, "A Vision of the Food System, 2045 CE—*re-set, rethink, re-do*";
- Professor Michael Bourlakis, Cranfield University, UK, "Blockchain and SC 4.0 in agri-food supply chains: Challenges and Opportunities";
- Professor Liudmyla Chip, Poltava State Agrarian University, Ukraine, "Features and prospects of Industry 4.0 in the Agrarian sector of Ukraine in wartime: economic and accounting aspects";
- Professor Olena Kopishynska, Poltava State Agrarian University, Ukraine, "Application of modern ERP systems for Agri-food Supply Chains as a strategy for reaching the level of technology 4.0 for non-manufacturing organizations".

The scientific program of the conference consisted of four themed plenary sessions across the two days alongside breaks for poster presentations and capacity building meetings, which were designed to stimulate further discussion and collaboration. Research abstracts were organized by theme and presented either in-person or online via Teams using a Logitech Rally Bar Mini All-in-One Video Conferencing System for authors of accepted abstracts who were unable to attend in person.

Finally, we would like to express our gratitude to Research England for the grant administered by Universities UK International, which made this conference possible. We would also like to thank the conference's Organizing Committee and Scientific Committee, as well as both the University of Leicester, UK, and Poltava State Agrarian University, Ukraine, as partners under the UK-Ukraine R&I twinning grants scheme. Special thanks go to the conference chairs, session chairs, reviewers, and volunteers for their efforts in making this conference a success, which was no small feat within the 5-month grant window.

Several Special Issues are collecting full-length papers based on the submitted abstracts:

- *Discover Food* (Springer) Special Issue: "The Role of Artificial Intelligence (AI) in Food Security" https://link.springer.com/collections/jaibdfaibi;
- *Sustainability* (MDPI) Special Issue: "Sustainable Supply Chain and Lean Manufacturing" https://www.mdpi.com/journal/sustainability/special_issues/5HY24R6111;
- *International Journal of Industrial Engineering and Operations Management* (IJIEOM) (Emerald) special issue "Digitizing Food Supply Chains: A Path to Ensuring Food Security" https://www.emeraldgrouppublishing.com/calls-for-papers/digitizing-food-supply-chains-a-path-ensuring-food-security;
- *International Journal of Food Science & Technology (IJFST) (Wiley) Special Issue: "Digital Technology to Combat Food Fraud in the Agri-Food Supply Chain" https://ifst.onlinelibrary.wiley.com/hub/journal/13652621/cfp_digital_technology_combat_food_fraud_agri_food_supply_chain*;
- *International Journal of Logistics Management* (IJLM) (Emerald) Special Issue: "Transforming Food Supply Chains: Harnessing the Potential of the Digital Era" https://www.emeraldgrouppublishing.com/calls-for-papers/transforming-food-supply-chains-harnessing-potential-digital-era.

We extend warm thanks to Ms. Rosalie Sun for the kind help and support with the editorial process for IC4AFSC 2023.

Conflicts of Interest: The authors declare no conflict of interest.

engineering proceedings

MDPI

Editorial

Statement of Peer Review †

Hana Trollman [1,*] and Iuliia Samoilyk [2]

1 School of Business, University of Leicester, Leicester LE2 1RQ, UK
2 Economy and International Economic Relations Department, Poltava State Agrarian University, 36003 Poltava, Ukraine
* Correspondence: ht203@leicester.ac.uk
† Presented at the International Conference on Industry 4.0 for Agri-food Supply Chains: Addressing Socio-economic and Environmental Challenges in Ukraine, Leicester, UK and Online, 24–25 July 2023.

In submitting conference proceedings to *Engineering Proceedings*, the volume editors of the proceedings certify to the publisher that all papers published in this volume have been subjected to peer review administered by the volume editors. Reviews were conducted by expert referees to the professional and scientific standards expected of a proceedings journal.

- Type of peer review: double-blind
- Conference submission management system: Excel
- Number of submissions sent for review: 35
- Number of submissions accepted: 27
- Acceptance rate (number of submissions accepted/number of submissions received): 77.1%
- Average number of reviews per paper: 2
- Total number of reviewers involved: 26
- Any additional information on the review process: None.

Conflicts of Interest: The authors declare no conflict of interest.

Citation: Trollman, H.; Samoilyk, I. Statement of Peer Review. *Eng. Proc.* **2023**, *40*, 29. https://doi.org/10.3390/engproc2023040029

Published: 17 August 2023

engineering proceedings

Abstract

Revolutionizing Food Safety in Agri-Food Supply Chains: The Impact of Autonomous Mobile Robots on Inspections †

Abhay K. Grover

Robert H. Smith School of Business, University of Maryland, College Park, MD 20742, USA; akgrover@umd.edu

† Presented at the International Conference on Industry 4.0 for Agri-food Supply Chains: Addressing Socio-economic and Environmental Challenges in Ukraine, Leicester, UK and Online, 24–25 July 2023.

Keywords: autonomous mobile robots; food inspections; food safety

Food safety managers across agri-food supply chains face complex challenges in meeting regulatory requirements, managing consumer expectations, and dealing with a shortage of qualified professionals. Despite efforts, food facilities still struggle with foundational safety issues. This study aims to explore how the deployment of autonomous mobile robots (AMRs) augments the efforts of food safety professionals by helping them to perform inspections.

Unlike conventional robots, AMRs can traverse complicated physical spaces without human aid and use wireless communication for real-time assessments, record-keeping, and regulatory compliance. AMRs come in different forms and have advanced sensors and AI-enabled processors. They can be trained manually and guided to relevant inspection points using a digital twin or 3D map. AMRs offer several benefits, including inspections in hazardous locations, sanitization capabilities, thermal and visual inspections, facility patrolling and security, and pest control. Challenges include operating in harsh conditions, mobility in complex environments, battery charging, and cost considerations.

Despite challenges, AMRs indicate promise in increasing process efficacy, lowering downtime, increasing safety, and permitting data analysis. This paper identifies the advantages and challenges of AMR adoption for food inspections using specific use cases.

Funding: This research received no external funding.

Institutional Review Board Statement: Not applicable.

Informed Consent Statement: Not applicable.

Data Availability Statement: Not applicable.

Conflicts of Interest: The author declares no conflict of interest.

Citation: Grover, A.K. Revolutionizing Food Safety in Agri-Food Supply Chains: The Impact of Autonomous Mobile Robots on Inspections. *Eng. Proc.* **2023**, *40*, 2. https://doi.org/10.3390/engproc2023040002

Academic Editor: Hana Trollman

Published: 14 July 2023

engineering proceedings

MDPI

Abstract

Analyzing the Barriers to Implementing Industry 4.0 for Enhanced Traceability in the Agri-Food Supply Chain †

Bishal Dey Sarkar * and **Isha Sharma**

Symbiosis Institute of Operations Management, Symbiosis International (Deemed University), Pune 412115, India; isha.sharma@siom.in

* Correspondence: bishal.sarkar@siom.in

† Presented at the International Conference on Industry 4.0 for Agri-food Supply Chains: Addressing Socio-economic and Environmental Challenges in Ukraine, Leicester, UK and Online, 24–25 July 2023.

Keywords: traceability; Industry 4.0; agri-food supply chain; fuzzy set theory; severity value

In recent times, Industry 4.0 has gained significant recognition due to its potential to revolutionize diverse sectors. It involves integrating intelligent technologies to enable digitalization within supply chains. The agri-food supply chain benefits from Industry 4.0 by providing avenues to enhance traceability and tackle the socio-economic and environmental challenges aligned with sustainable development goals. Implementing Industry 4.0 technologies, such as security measures, transparency enhancements, and traceability systems, can greatly enhance quality and environmental monitoring, precision agriculture, process automation, and more. Emphasizing the socio-economic perspective, Industry 4.0 contributes to the economic growth and competitiveness of the agri-food industry. These technologies foster consumer trust by improving traceability and transparency, leading to increased demand and improved market access for agricultural products. This, in turn, strengthens the revenue and profitability of agri-food businesses, facilitating economic development and job creation. Nonetheless, various barriers hinder the widespread adoption of Industry 4.0 in the agri-food supply chain.

Further, the literature highlights various research gaps. To close these gaps, this study develops various research questions: How can the agricultural sector leverage Industry 4.0 in the agri-food supply chain for implementing traceability? How can stakeholders in the agriculture sector effectively integrate Industry 4.0 into their operations? What barriers are there to the adoption of Industry 4.0 in the agri-food supply chain for implementing traceability? Which barriers must be prioritized for successfully adopting Industry 4.0 in the agri-food supply chain? How are the identified barriers related to each other? From the aforementioned research questions, the study formulates the following research objectives: (i) to identify the barriers to implementing Industry 4.0 for enhanced traceability in the agri-food supply chain; (ii) to obtain the severity values of the identified barriers; (iii) to rank the barriers based on their influence, and suggest strategies for the successful implementation of Industry 4.0 for enhanced traceability in the agri-food supply chain; and (iv) to investigate the interrelationships among the barriers. The study develops a framework by integrating various theories and methodologies based on the research objectives. It combines an evidential reasoning approach with fuzzy set theory, triple bottom line theory, and stakeholder theory.

This study explores these barriers in Ukraine's agri-food supply chain, specifically with the goal of enhancing traceability. By comprehending and addressing these challenges, the study aims to leverage Industry 4.0 technologies to improve traceability, promote transparency, and address key socio-economic and environmental concerns within Ukraine's agri-food industry. A comprehensive list of barriers was identified from the literature and further validated by the experts. The developed framework ranks the barriers based on

Citation: Sarkar, B.D.; Sharma, I. Analyzing the Barriers to Implementing Industry 4.0 for Enhanced Traceability in the Agri-Food Supply Chain. *Eng. Proc.* **2023**, *40*, 3. https://doi.org/10.3390/engproc2023040003

Academic Editor: Hana Trollman

Published: 17 July 2023

the severity values of various perspectives, and also provides the overall ranking of the barriers. The developed framework uses numeric and subjective data. Hence, quantitative and qualitative attributes can be modeled under volatile, uncertain, complex, and ambiguous environments. This helps to monitor, improve business, develop strategies, and achieve clarity and transparency in the overall system. The obtained hierarchy of the barriers helps decision makers identify significant areas that need more attention. The research findings will contribute to academic knowledge and offer valuable insights to practitioners, managers, and policymakers in Ukraine. These insights will assist them in navigating challenges and fostering a supportive environment for adopting Industry 4.0 solutions. Ultimately, this endeavor will improve traceability, transparency, and sustainability within Ukraine's agri-food industry, effectively addressing critical socio-economic and environmental concerns.

However, other barriers may be discovered in future studies. To study the changing behavior of the barriers, belief degrees could be used instead of changing the severity values. The study focuses on environmental, economic, and social dimensions to develop a hierarchy; other dimensions, like operational, technical, political, and legal, could be used to establish a hierarchy of barriers.

Author Contributions: Conceptualization, methodology, software, formal analysis, investigation was done by B.D.S., and original draft preparation, writing, review and editing was carried out by I.S. All authors have read and agreed to the published version of the manuscript.

Funding: This research received no external funding.

Institutional Review Board Statement: Not applicable.

Informed Consent Statement: Not applicable.

Data Availability Statement: Data sharing is not applicable to this article.

Conflicts of Interest: The authors declare no conflict of interest.

engineering proceedings

MDPI

Abstract

Food Security via Food Waste Prevention: Categorization of Household Food Waste for Artificial Intelligence-Driven Interventions †

Nurul Islam Mohamed Yusoff [1,*], Elliot Woolley [1] and Jan Godsell [2]

[1] Wolfson School of Mechanical, Electrical and Manufacturing Engineering, Loughborough University, Leicestershire LE11 3TU, UK; e.b.woolley@lboro.ac.uk

[2] Loughborough Business School, Loughborough University, Leicestershire LE11 3TU, UK; j.godsell@lboro.ac.uk

* Correspondence: n.i.mohamed-yusoff@lboro.ac.uk

† Presented at the International Conference on Industry 4.0 for Agri-Food Supply Chains: Addressing Socio-Economic and Environmental Challenges in Ukraine, Leicester, UK and Online, 24–25 July 2023.

Keywords: artificial intelligence; household food waste; food security; COM-B

Global food waste, around 1.3 billion tons yearly, presents significant implications, notably aggravating food insecurity for approximately 768 million people. Households contribute 61% of this waste, underlining the potential to increase food availability. Despite the significance of household food waste (HFW), the lack of standardized categorization poses challenges in addressing this issue effectively. This study establishes a comprehensive framework for understanding HFW within the food supply chain, aiming to enable targeted interventions to reduce specific waste categories. A systematic review of existing food waste and categorization literature is conducted to provide a background on the global food waste problem and the specific reasons why household food waste occurs. A framework that implements a nine-stage categorization process for HFW (see Figure 1) is described, resulting in 37 distinct food waste categories stemming from domestic environments. Results indicate that a significant proportion of HFW consists of edible and avoidable food waste, emphasizing the need for interventions that target these specific waste streams. By utilizing this framework, much-improved clarity about the types of food waste could be gained from consumer-focused studies. This improved data granularity would then enable far more targeted interventions to be conceived and developed to address the most prolific or problematic (e.g., highest cost or environmental impact) HFW streams. Due to increased data requirements, we proposed and discussed the suitability of this framework for implementation using intelligent systems. And the current research is discussed in the context of AI-driven customized food waste prevention capabilities. Ongoing work is addressing structured reasoning for household food waste according to household food management stages and the well-known COM-B model (capability, opportunity, motivation, behavior change). In conclusion, this study emphasizes the need for a standardized categorization to understand HFW and design effective data-driven, AI-assisted interventions based on the COM-B behavioral change model. By addressing the categorization challenges and offering practical recommendations, this study provides valuable insights for policymakers, researchers, and practitioners to reduce HFW, promote sustainable consumption practices, and increase global food security.

Citation: Yusoff, N.I.M.; Woolley, E.; Godsell, J. Food Security via Food Waste Prevention: Categorization of Household Food Waste for Artificial Intelligence-Driven Interventions. *Eng. Proc.* **2023**, *40*, 4. https://doi.org/10.3390/engproc2023040004

Academic Editor: Hana Trollman

Published: 14 July 2023

1	Edibility	
	Edible	Inedible
2	Dietry Requirement	
	Required by diets	Over consumptions
3	State	
	Eatable	Not Eatable
	Not eatable by humans, eatable by animals	
4	Inedible conversion	
	Edible if processed	Never edible
	Not edible for humans, edible for animals	
5	Avoidability	
	Avoidable	Potentially avoidable
6	Viable for conversion	
	Economically, socially, culturally acceptable	
	Not viable	
7	Origin	
	Animal-based	Plant-based
8	Animal product presence	
	Meat	Animal product
	Not in contact with or mixed with meat, animal products or by products	In contact with or mixed with meat, animal products or by products
9	Treatment	
	Cooked	Uncooked

Figure 1. Indicators to categorize food waste.

Author Contributions: Conceptualization, N.I.M.Y., E.W. and J.G.; methodology, N.I.M.Y. and E.W.; formal analysis, N.I.M.Y. and E.W.; resources, N.I.M.Y., E.W. and J.G.; data curation, N.I.M.Y. and E.W.; writing—original draft preparation, N.I.M.Y.; writing—review and editing, N.I.M.Y.; visualization, N.I.M.Y. and E.W.; supervision, E.W. and J.G.; project administration, N.I.M.Y., E.W.; funding acquisition, N.I.M.Y. All authors have read and agreed to the published version of the manuscript.

Funding: This research was funded by Majlis Amanah Rakyat (REF NO: 330408389954).

Institutional Review Board Statement: Not applicable.

Informed Consent Statement: Not applicable.

Data Availability Statement: Not applicable.

Conflicts of Interest: The authors declare no conflict of interest.

engineering proceedings

MDPI

Abstract

Assessing Adoption Challenges of Blockchain Technology in Agri-Food Supply Chain †

Neha Gupta [1,*], Sumit Gupta [2,*] and Rahul Gupta [3]

1 Amity School of Business, Amity University Uttar Pradesh, Noida 201313, India
2 Department of Mechanical Engineering, Amity School of Engineering and Technology, Amity University Uttar Pradesh, Noida 201301, India
3 Amity Business School, Amity University Uttar Pradesh, Noida 201313, India; rgupta10@amity.edu
* Correspondence: ngngupta4@gmail.com (N.G.); sgupta20@amity.edu (S.G.)
† Presented at the International Conference on Industry 4.0 for Agri-food Supply Chains: Addressing Socio-Economic and Environmental Challenges in Ukraine, Leicester, UK and Online, 24–25 July 2023.

Abstract: The Agri-Food Supply Chain (AFSC) faces several barriers in developing economies, hindering the adoption of blockchain technology. However, the adoption of these technologies can transform traditional supply chains by incorporating transparency, traceability, transaction records, and enhanced security. This paper aims to evaluate the barriers impacting the adoption of blockchain in the AFSC and provide insights to improve performance and facilitate their effective implementation in emerging economies. Therefore, a comprehensive literature survey was conducted to identify the barriers, which were further validated through experts' opinions. The integrated interpretive structural modeling (ISM) and decision-making trial and evaluation laboratory (DEMATEL) technique were then employed to develop a structural model and determine the cause–effect relationships between the identified barriers. The study revealed several key barriers affecting the adoption of blockchain technology in the AFSC of developing economies. These barriers were analyzed using the ISM and DEMATEL technique, which provided insights into their interdependencies and impact on each other. The results offer a comprehensive understanding of the barriers and their causal relationships. This study provides unique insights for the agri-food sector to improve performance by addressing the identified barriers. The results can guide agri-food managers, blockchain technology service providers, and the government in formulating strategies and policies to effectively adopt blockchain technologies in the AFSC.

Keywords: blockchain technology; Agri-Food Supply Chain (AFSC); developing economies; barriers; ISM; DEMATEL

Citation: Gupta, N.; Gupta, S.; Gupta, R. Assessing Adoption Challenges of Blockchain Technology in Agri-Food Supply Chain. *Eng. Proc.* **2023**, *40*, 5. https://doi.org/10.3390/engproc2023040005

Academic Editor: Hana Trollman

Published: 14 July 2023

Author Contributions: Conceptualization, N.G., S.G. and R.G.; methodology, N.G. and S.G.; software, N.G. and S.G.; validation, N.G., S.G. and R.G.; formal analysis, N.G. and S.G.; investigation, N.G. and S.G.; resources, S.G.; data curation, N.G. and S.G.; writing—original draft preparation, N.G. and R.G.; writing—review and editing, S.G.; visualization, S.G.; supervision, S.G.; project administration, S.G. All authors have read and agreed to the published version of the manuscript.

Funding: This research received no external funding.

Institutional Review Board Statement: Not applicable.

Informed Consent Statement: Not applicable.

Data Availability Statement: Data available on request due to privacy/ethical restrictions.

Conflicts of Interest: The authors declare no conflict of interest.

engineering proceedings

Abstract

The Challenges and Opportunities for the Development of Industry 4.0 and Agri-Food Supply Chain in the Context of Energy Infrastructure Restoration in Ukraine †

Lyudmyla Svystun [1,*], Iuliia Samoilyk [2] and Mykola Svystun [2]

1 Public Organization Sustainable Development Platform "Perspektiva", 36000 Poltava, Ukraine
2 Department of Economics and International Economic Relations, Poltava State Agrarian University, 36003 Poltava, Ukraine; iuliia.samoilyk@gmail.com (I.S.); niksv@gmail.com (M.S.)
* Correspondence: svmila308@gmail.com

† Presented at the International Conference on Industry 4.0 for Agri-food Supply Chains: Addressing Socio-economic and Environmental Challenges in Ukraine, Leicester, UK and Online, 24–25 July 2023.

Keywords: energy infrastructure; agri-food supply chain; alternative energy; hydrogen energy; Industry 4.0; reconstruction; rural areas

The war in Ukraine is one of the most catastrophic events of the 21st century, and it has been accompanied by significant civilian casualties and the destruction of buildings, structures, and infrastructure. The Ukrainian economy has experienced a loss of 50–60% of its "unproduced" gross domestic product. Revenue to the state budget has decreased by 70% from customs authorities and by 30% from tax authorities. The reconstruction of Ukraine's national economy will require active development of the energy sector and the involvement of foreign companies to restore the social and economic infrastructure. The issue of energy development is closely linked to the formation of food chains, as the harvest and price of agricultural products depend on the cost and availability of energy resources.

Ukraine faces the challenge of centralizing and unifying large-scale power-generation facilities, such as nuclear power plants, thermal power plants, hydroelectric power plants, etc. The deployment of Russian military aggression against Ukraine and constant missile attacks on energy infrastructure objects pose significant challenges to the functioning of the power supply system. For instance, as of 10 February 2023, nearly twenty thermal power plant units remain damaged due to constant Russian attacks. Considering the occupation of certain energy facilities, Ukraine has temporarily lost 44% of its nuclear generation capacity, 75% of its thermal power plants, and 33% of its block heating power plants. Energy infrastructure in conflict-affected areas has been almost completely destroyed and requires restoration based on modern and innovative foundations. The development of alternative energy and energy generation methods in liberated communities could serve as a solution. Given Ukraine's energy dependence on supplies and the constant increase in energy prices, changing its energy strategy by implementing effective energy conservation programs and developing alternative energy sources, including in the agricultural sector, is of strategic importance and requires urgent attention. Ukraine's Energy Strategy until 2030 recognizes the exploration of alternative energy sources as a crucial factor in increasing energy security, particularly in light of Ukraine's European integration aspirations.

However, little attention has been given to the technology involved in energy generation through the use of solar energy, as well as the storage and accumulation of hydrogen gas. These technologies are mainstream aspects of Industry 4.0. Similar facilities are already being used in Australia and some European countries, primarily in the private sector, enabling households to meet their energy needs and redirect surplus energy for sale. Currently, a significant portion of Ukraine's population resides in private homes in rural

Citation: Svystun, L.; Samoilyk, I.; Svystun, M. The Challenges and Opportunities for the Development of Industry 4.0 and Agri-Food Supply Chain in the Context of Energy Infrastructure Restoration in Ukraine. *Eng. Proc.* **2023**, *40*, 6. https://doi.org/10.3390/engproc2023040006

Academic Editor: Hana Trollman

Published: 17 July 2023

areas (over 40%). The same situation applies to the currently occupied territories. Additionally, nearly all internally displaced people have lost their jobs and income. Therefore, establishing energy generation systems in private households in de-occupied rural areas would provide the opportunity to achieve the following goals: meet the basic needs of the population, ensure stable incomes for households, strengthen community autonomy and Ukraine's energy independence, foster the development of modern ecological technologies in energy and transportation, and prevent agri-food production from decreasing.

Author Contributions: Conceptualization, M.S.; methodology, L.S.; formal analysis, M.S.; investigation, M.S.; resources, I.S.; writing—original draft preparation, L.S.; writing—review and editing, L.S.; visualization, I.S.; project administration, I.S.; funding acquisition, I.S. All authors have read and agreed to the published version of the manuscript.

Funding: This research was funded by UK Research and Innovation (UKRI).

Institutional Review Board Statement: Not applicable.

Informed Consent Statement: Not applicable.

Data Availability Statement: Publicly available datasets were analyzed in this study. This data can be found here: https://business.diia.gov.ua/wartime (accessed on 1 June 2023), https://ces.org.ua (accessed on 1 June 2023), https://zakon.rada.gov.ua/signal/kr06145a.doc (accessed on 1 June 2023), http://ive.org.ua/ (accessed on 1 June 2023).

Conflicts of Interest: The authors declare no conflict of interest.

engineering proceedings

MDPI

Abstract

Framework for Assessing Trust in the Use of Blockchain Technology in Agrifood Supply Chains †

Vaibhav Sharma, Anbesh Jamwal and Rajeev Agrawal *

Department of Mechanical Engineering, Malaviya National Institute of Technology Jaipur, Jaipur 302017, India; vaib.me81@gmail.com (V.S.); anveshjamwal73@gmail.com (A.J.)
* Correspondence: ragrawal.mech@mnit.ac.in
† Presented at the International Conference on Industry 4.0 for Agri-food Supply Chains: Addressing Socio-economic and Environmental Challenges in Ukraine, Leicester, UK and Online, 24–25 July 2023.

Abstract: The advent of Industry 4.0 (I4.0) technologies has revolutionized production environments with their application in supply chains, particularly within the agrifood sector. One notable I4.0 technology is blockchain, which holds significant potential for traceability in agrifood supply chains. However, there are concerns regarding digital trust among the actors involved in adopting this technology. The concept of digital trust, essential for successful implementation, remains underexplored. This research aims to propose a framework for evaluating digital trust in the context of blockchain technology to foster a secure and reliable information sharing environment among all stakeholders within the agrifood supply chain to build confidence in security based on data permissions for user identity. To accomplish this, an extensive literature review was conducted to identify the factors affecting stakeholders' expectations and trust in using blockchain technology in agrifood supply chains. The literature review will enhance the knowledge about these different factors affecting digital trust under four key dimensions, that are, security/privacy, data control, accountability, and benefit/value. These factors are then ranked using a multi criteria decision-making technique, enabling the development of a framework for industries and government organizations. This framework addresses the use of blockchain technology for traceability in agrifood supply chains while ensuring the trust of actors utilizing this technology. In regions facing war-like situations, such as Ukraine, it becomes crucial to evaluate the factors that can enhance food safety in agrifood supply chains, mitigate food waste and fraud risks, and maintain the supply chain sustainability by exploring alternative food supplies from reliable partners. The situation is the same all over the globe, in which supply chain risks include frauds and lack of transparency. This study outlines the managerial implications and suggests future research directions to develop a model for assessing digital trust. This model aims to foster information sharing among actors, considering aspects, such as willingness, vulnerability acceptance, shared values, security, identifiability, and digital trust. Smart contracts can be added to the model that removes the need for a third party, warranting more trust. The deployment of this model on a private or public blockchain can enhance transparency, traceability, and address food safety concerns within transactions by addressing issues of security and reliability, accountability, and oversight with an inclusive, ethical, and responsible use.

Keywords: digital trust; agrifood supply chain; traceability; food safety

Citation: Sharma, V.; Jamwal, A.; Agrawal, R. Framework for Assessing Trust in the Use of Blockchain Technology in Agrifood Supply Chains. *Eng. Proc.* **2023**, *40*, 7. https://doi.org/10.3390/engproc2023040007

Academic Editor: Hana Trollman

Published: 17 July 2023

Author Contributions: Conceptualization, V.S.; methodology, V.S. and A.J.; software, V.S. and A.J.; validation, R.A.; formal analysis, R.A.; investigation, R.A.; resources, R.A.; data curation, R.A.; writing—original draft preparation, V.S.; writing—review and editing, R.A.; visualization, V.S.; supervision, R.A.; project administration, R.A. All authors have read and agreed to the published version of the manuscript.

Funding: This research received no external funding.

Institutional Review Board Statement: Not applicable.

Informed Consent Statement: Not applicable.

Data Availability Statement: Not applicable.

Conflicts of Interest: The authors declare no conflict of interest.

engineering proceedings

MDPI

Abstract

Building Resilience and Overcoming Challenges in War Times: The Role of Lean 4.0 in Agri-Food Supply Chains †

Michelle Grace Tetteh *, Sandeep Jagtap and Konstantinos Salonitis

Sustainable Manufacturing Systems Centre, School of Aerospace Transport and Manufacturing, Cranfield University, Cranfield MK43 0AL, UK; s.z.jagtap@cranfield.ac.uk (S.J.); k.salonitis@cranfield.ac.uk (K.S.)

* Correspondence: michelle.tetteh@cranfield.ac.uk

† Presented at the International Conference on Industry 4.0 for Agri-food Supply Chains: Addressing Socio-economic and Environmental Challenges in Ukraine, Leicester, UK and Online, 24–25 July 2023.

Keywords: Industry 4.0; Lean 4.0; agri-food supply chain; Russia-Ukraine; food resilience

The ongoing conflict and the impact of Russia–Ukraine war pose significant challenges to various sectors, including the agri-food industry. The agricultural industry serves as a backbone for many world economies, contributing significantly to their gross domestic product. The continuing war and shifting economic environment have greatly affected grain production, encompassing crops like wheat, maize, and barley. Farmlands have been destroyed, output has been reduced, and transportation and export lines have been disrupted, resulting in a substantial impact on the grain business. Conflict-affected areas are witnessing the negative consequences of disrupted supply chains, the emergence of animal diseases, and reduced access to feed and veterinary services, thereby harming livestock and dairy farms. The horticulture sector, specializing in fruit and vegetables, has also faced setbacks due to reduced investments, damaged infrastructure, and limited access to export markets. Furthermore, the war has affected the production, processing, and export activities of wheat and major oilseed crops such as sunflower, soybean, and rapeseed.

While the war has undeniably influenced agri-food sectors, it is crucial to emphasize the potential combination of Industry 4.0 technologies and Lean strategies (Lean 4.0) in building resilience and overcoming various challenges. By adopting Lean 4.0, these sectors can achieve superior productivity, increased overall efficiency, reduced waste, and enhanced transparency. This research investigates the potential of Lean 4.0 in revitalizing the agri-food sector, specifically addressing the challenges and strategies for its current implementation in the post-war context.

Citation: Tetteh, M.G.; Jagtap, S.; Salonitis, K. Building Resilience and Overcoming Challenges in War Times: The Role of Lean 4.0 in Agri-Food Supply Chains. *Eng. Proc.* **2023**, *40*, 8. https://doi.org/10.3390/engproc2023040008

Academic Editor: Hana Trollman

Published: 17 July 2023

Author Contributions: Conceptualization, M.G.T. and S.J.; methodology, M.G.T.; validation, M.G.T., S.J. and K.S.; formal analysis, M.G.T.; investigation, M.G.T.; resources, S.J.; data curation, M.G.T.; writing—original draft preparation, M.G.T., S.J. and K.S.; writing—review and editing, M.G.T., S.J. and K.S.; visualization, M.G.T.; supervision, S.J.; project administration, K.S.; funding acquisition, S.J. All authors have read and agreed to the published version of the manuscript.

Funding: This research received no external funding.

Institutional Review Board Statement: Not applicable.

Informed Consent Statement: Not applicable.

Data Availability Statement: Not applicable.

Conflicts of Interest: The authors declare no conflict of interest.

engineering proceedings

MDPI

Abstract

Analyzing Sustainability Assessment Factors Influencing Agriculture Supply Chains in the Age of Industry 4.0 †

Harshad Sonar [1,*], Nikhil Ghag [2] and Isha Sharma [1]

1 Symbiosis Institute of Operations Management (SIOM), Symbiosis International (Deemed) University, Nashik 422008, India; isha.sharma@siom.in
2 National Institute of Industrial Engineering (NITIE), Powai, Mumbai 400087, India; nikhilghg3@gmail.com
* Correspondence: harshad.sonar@siom.in or harshadsonar@gmail.com

† Presented at the International Conference on Industry 4.0 for Agri-food Supply Chains: Addressing Socio-economic and Environmental Challenges in Ukraine, Leicester, UK and Online, 24–25 July 2023.

Keywords: Industry 4.0; sustainability; agriculture supply chains; MCDM

Industry 4.0 is paving the way toward a new age of industrial revolution, wherein new-age technologies have a humungous impact in revamping supply chains. The invention of AI, big data, and the internet of things is reshaping supply chains into highly intelligent, rational, and interconnected networks capable of prudent decision-making. The incorporation of new-age technologies has revolutionized real-time and seamless data flow and exchange, apprehensive and foresighted analysis, and coherent stakeholder collision. This transformation enables businesses to achieve complete visibility across their processes, optimize inventory control, improve product traceability, and increase satisfaction among consumers. By adopting these emerging technologies, businesses can unlock new windows of opportunities, surmount impediments, and achieve sustainable and competitive advantage in the dynamic world of supply chain management.

The agriculture sector, which is one of the oldest and most important sectors of the global economy, has also started to embrace Industry 4.0 technologies to improve its supply chain management. The agriculture industry faces a number of difficulties, such as the need for sustainable production methods, rising food consumption, and climate change. By utilizing cutting-edge technologies to boost productivity, efficiency, and sustainability in agriculture supply chains, Industry 4.0 offers a special chance to address these issues. Because Industry 4.0 technologies have the potential to revolutionize food production and distribution, they are becoming increasingly important for supply chain management in agriculture. Industry 4.0 implementation in the agri-food sector has the potential to improve inter-organizational technological systems, integrated planning and execution systems, real-time visibility, agri-good procurement, and agri-good inventory management. The agri-food sector may enhance the integration and coordination of technological systems between organizations by utilizing cutting-edge technologies like IoT, AI, and big data analytics. Despite these technologies being available and the benefits of incorporating Industry 4.0 into agriculture supply chains, the agriculture sector is facing lots of issues.

The existing literature does not offer sufficient empirical evidence regarding the implementation of Industry 4.0 technologies in agriculture supply chains. Agriculture supply chains (ASCs) appear to be deprived of this transformational reform in terms of implementation and research. Moreover, current research also overlooks various sustainability assessment factors that hinder the adoption of Industry 4.0 in the agriculture sector. This study intends to identify the sustainability assessment factors of Industry 4.0 in agriculture supply chains. This study utilizes fuzzy-Delphi and DEMATEL methodologyto evaluate the interdependencies between the identified factors.

Citation: Sonar, H.; Ghag, N.; Sharma, I. Analyzing Sustainability Assessment Factors Influencing Agriculture Supply Chains in the Age of Industry 4.0. *Eng. Proc.* **2023**, *40*, 9. https://doi.org/10.3390/engproc2023040009

Academic Editor: Hana Trollman

Published: 17 July 2023

Fuzzy-Delphi method is used to identify and finalize the factors based on the expert opinions. DEMATEL is a technique for analyzing and viewing a structural model with intricate causal relationships. Studies on supplier choice, green supply chain management, business process management, and hospitality have all made use of DEMATEL. This study discusses the DEMATEL approach, which can analyze the interdependencies between sustainability assessment elements for integrating Industry 4.0 technologies into agriculture supply chains and determine the performance effects of these aspects. Finally, this study integrates these two approaches to offer a framework for agriculture supply chain businesses and suggests a novel approach for the sustainable evaluation of Industry 4.0 technologies. Eleven specialists with a minimum of five years of expertise in their respective disciplines provided the information.

The results of this study will enable managers, practitioners, and planners to mitigate sustainability assessment factors related to Industry 4.0 more successfully, thereby promoting sustainability in the agriculture supply chains of developing nations. In terms of policy, this study can assist institutions of a particular area in understanding the degree of digitalization of current agri-cooperatives and, as a result, in better designing political instruments to promote a more sustainable environment.

Author Contributions: Conceptualization, H.S.; methodology, N.G.; investigation, N.G.; data curation, I.S.; writing—original draft preparation, N.G. and I.S.; writing—review and editing, H.S. All authors have read and agreed to the published version of the manuscript.

Funding: This research received no external funding.

Institutional Review Board Statement: Not applicable.

Informed Consent Statement: Not applicable.

Data Availability Statement: Not applicable.

Conflicts of Interest: The authors declare no conflict of interest.

engineering proceedings

MDPI

Abstract

Revolutionizing Food Production: Integrating Industry 4.0 and Carbon Neutral Strategies for a Sustainable Future †

Noor Imran *, Sandeep Jagtap, Mohamed Afy-Shararah and Konstantinos Salonitis

Sustainable Manufacturing Systems Centre, School of Aerospace Transport and Manufacturing, Cranfield University, Cranfield MK43 0AL, UK; s.z.jagtap@cranfield.ac.uk (S.J.); m.a.shararah@cranfield.ac.uk (M.A.-S.); k.salonitis@cranfield.ac.uk (K.S.)

* Correspondence: noor.imran.596@cranfield.ac.uk

† Presented at the International Conference on Industry 4.0 for Agri-Food Supply Chains: Addressing Socio-Economic and Environmental Challenges in Ukraine, Leicester, UK and Online, 24–25 July 2023.

Keywords: carbon neutral strategies in the food industry; circular economy for food; Life Cycle Assessment (LCA) in food production; Industry 4.0 in food manufacturing; carbon offsetting in the food sector

The food industry is actively adopting carbon neutral strategies aligned with the principles of Industry 4.0 to establish long-term goals for reducing carbon emissions and accelerating the transition to a low-carbon food production industry. Industry 4.0, often referred to as the fourth industrial revolution, emphasizes the integration of digital technologies, automation, and data exchange into manufacturing processes.

In the context of the food industry, Industry 4.0 technologies offer significant potential to optimize production processes, improve energy efficiencies, and reduce greenhouse gas emissions. This study explores the concept of a "circular economy" for food, which aligns with the principles of Industry 4.0 by promoting resource efficiency, waste reduction, and the reuse of by-products.

To assess the environmental impact of food manufacturing operations, the research methodology incorporates the theory and methods of a Life Cycle Assessment (LCA), a commonly used tool in Industry 4.0 practices for analyzing the environmental performance of products and processes throughout their entire life cycle.

The data collection process involves gathering qualitative and quantitative data on various parameters, including energy consumption, energy efficiency, packaging materials, water usage, waste generation, and waste management practices. By integrating Industry 4.0 technologies, the study aims to optimize the production life cycle and reduce the carbon footprint of food manufacturing operations.

The results of this research can contribute to a more sustainable food sector by enabling a reduction in carbon emissions. Through utilization of Industry 4.0 principles, renewable energy sources, and eco-friendly packaging strategies, carbon offsetting becomes attainable, allowing the food industry to meet emission reduction targets. Furthermore, the study highlights that carbon offset projects not only reduce emissions but also deliver additional environmental, social, and economic benefits.

In conclusion, the proposed solutions integrate Industry 4.0 principles with carbon offsetting strategies, fostering a paradigm shift among food manufacturers. These strategies should be tailored to the specific needs and financial situations of companies, ensuring their feasibility and broad industry benefits. By embracing the opportunities offered by Industry 4.0 and implementing these solutions, the food industry can make significant contributions to reducing carbon emissions and achieving a more sustainable future.

Citation: Imran, N.; Jagtap, S.; Afy-Shararah, M.; Salonitis, K. Revolutionizing Food Production: Integrating Industry 4.0 and Carbon Neutral Strategies for a Sustainable Future. *Eng. Proc.* **2023**, *40*, 10. https://doi.org/10.3390/engproc2023040010

Academic Editor: Hana Trollman

Published: 17 July 2023

Author Contributions: Conceptualization, N.I., S.J., and M.A.-S.; Methodology, N.I., and S.J.; validation, S.J., M.A.-S., and K.S.; writing—original draft preparation, N.I.; writing—review and editing, N.I. and S.J.; supervision, S.J. and M.A.-S. All authors have read and agreed to the published version of the manuscript.

Funding: This Research received no external funding.

Institutional Review Board Statement: Not applicable.

Informed Consent Statement: Not applicable.

Data Availability Statement: Not applicable.

Conflicts of Interest: The authors declare no conflict of interest.

engineering proceedings

Abstract

Safeguarding Food Industry: Understanding Cyberthreats and Ensuring Cybersecurity †

Adel Alqudhaibi *, Ashish Krishna, Sandeep Jagtap, Mohamed Afy-Shararah and Konstantinos Salonitis

Sustainable Manufacturing Systems Centre, School of Aerospace Transport and Manufacturing, Cranfield University, Cranfield MK43 0AL, UK; ashish.krishna.831@cranfield.ac.uk (A.K.); s.z.jagtap@cranfield.ac.uk (S.J.); m.a.shararah@cranfield.ac.uk (M.A.-S.); k.salonitis@cranfield.ac.uk (K.S.)

* Correspondence: adel.alqudhaibi@cranfield.ac.uk

† Presented at the International Conference on Industry 4.0 for Agri-food Supply Chains: Addressing Socio-economic and Environmental Challenges in Ukraine, Leicester, UK and Online, 24–25 July 2023.

Abstract: The food industry stands as one of the most vital manufacturing sectors globally, with an ever-increasing reliance on digitalization and technology-driven processes. However, this advancement comes with an inherent risk of cyberattacks, encompassing data breaches and system disruptions, which can severely impact production and disrupt the entire food supply chain. Consequently, such cyberthreats can lead to consumer fear and mistrust, potentially tarnishing a company's brand image. Additionally, the sector is becoming the focus of cyberthreat actors owing to the current crisis in Ukraine, revealing the severity of the rippling effects of these disruptions. This research aims to delve into the current perception of cyberthreats within the food industry, emphasizing the importance of cybersecurity and analyzing the measures taken by stakeholders to mitigate the risks associated with cyberattacks. The findings reveal that although the food industry acknowledges the potential threats posed by inadequate cybersecurity measures, these risks are perceived as insignificant due to the unique nature of the industry. Moreover, an extensive literature review highlights that the food industry places great emphasis on adopting innovative information technologies to enhance operational efficiency and cost-effectiveness. However, it remains vulnerable to cyberattacks, necessitating continuous employee education and training to strengthen the security landscape. This holistic approach fosters a seamless, reliable, and sustainable growth environment for the industry. By analyzing the existing challenges and requirements, this study underscores the need for proactive measures to safeguard the food industry against cyberthreats. It emphasizes the significance of implementing robust cybersecurity protocols and cultivating a culture of awareness and preparedness within organizations. Furthermore, the research emphasizes the importance of employee education and training, equipping them with the necessary knowledge and skills to identify and mitigate potential cyber risks. In conclusion, while cognizant of the risks posed by cyberattacks, the food industry must prioritize cybersecurity measures to protect its production and supply chain. Enhancing the security environment through ongoing employee education and training is crucial for fostering consumer trust and enabling seamless growth within the industry. By adopting a proactive approach to cybersecurity, the food industry can ensure the sustainability and reliability of its operations in the face of evolving cyberthreats.

Keywords: cybersecurity in the food industry; cyberthreats and food production; food industry and cyberattacks; preventing cyberattacks in food distribution; importance of employee education in food industry cybersecurity

Citation: Alqudhaibi, A.; Krishna, A.; Jagtap, S.; Afy-Shararah, M.; Salonitis, K. Safeguarding Food Industry: Understanding Cyberthreats and Ensuring Cybersecurity. *Eng. Proc.* **2023**, *40*, 11. https://doi.org/10.3390/engproc2023040011

Academic Editor: Hana Trollman

Published: 17 July 2023

Author Contributions: Data curation, A.A. and A.K.; Formal analysis, A.A., A.K., S.J., M.A.-S. and K.S.; validation, A.A. and S.J.; investigation, A.A., A.K., S.J. and M.A.-S.; methodology, A.A., A.K., S.J. and M.A.-S.; Writing—original draft, A.A., A.K., S.J., M.A.-S. and K.S.; writing—review and editing, A.A., A.K., S.J., M.A.-S. and K.S.; visualization, S.J. and K.S.; supervision, S.J.; software, A.A., S.J. and M.A.-S.; resources, M.A.-S.; project administration, K.S. All authors have read and agreed to the published version of the manuscript.

Funding: This research received no external funding.

Institutional Review Board Statement: Not applicable.

Informed Consent Statement: Not applicable.

Data Availability Statement: Not applicable.

Conflicts of Interest: The authors declare that they have no known competing financial interest or personal relationships that could have appeared to influence the work reported in this paper.

engineering proceedings

MDPI

Abstract

Intelligent Assessment of Reusable Plastic Food Packaging for a Circular Supply Chain †

Mariam Qaiser [1,*], Elliot Woolley [1], James Colwill [1], Fiona L. Hatton [2] and D. Patrick Webb [1]

1 Wolfson School of Mechanical, Electrical and Manufacturing Engineering, Loughborough University, Loughborough LE11 3TU, UK; e.b.woolley@lboro.ac.uk (E.W.); j.a.colwill@lboro.ac.uk (J.C.); d.p.webb@lboro.ac.uk (D.P.W.)
2 Department of Materials, Loughborough University, Loughborough LE11 3TU, UK; f.hatton@lboro.ac.uk
* Correspondence: m.qaiser@lboro.ac.uk

† Presented at the International Conference on Industry 4.0 for Agri-food Supply Chains: Addressing Socio-Economic and Environmental Challenges in Ukraine, Leicester, UK and Online, 24–25 July 2023.

Keywords: sustainable manufacturing; environmental impacts; circular economy; polymers; waste reduction

The agri-food supply chain uses plastic food packaging alongside many other types of packaging materials. The extensive use of single use plastic packaging is known to have many significant negative environmental implications and efforts are being made to address this issue. One of the solutions for reducing the impacts of plastic packaging is to take a circular economy approach, which includes multiple reuses of packaging before closed-loop recycling. For this purpose, it is very important to investigate the factors that may influence, discourage or prevent reusability. This research focuses on the most concerning issue relating to the reuse of food packaging: the prevention of contamination due to crossover from the old product to the new. Specifically, this work investigates factors related to surface properties that have an influence on the effectiveness of cleaning plastic food packaging and on the post-cleaning assessment.

In this study, we examine the surface characteristics of recycled polyethylene terephthalate (rPET) trays that will likely change due to several supply use–clean cycles. Standard rPET packs (W × L × H, 150 × 210 × 40 mm) have been subjected to repeated wash cycles (using a Classeq® Glasswasher G400 Duo, Nisbets, Bristol, UK) at 55 °C wash and 70 °C sanitisation, and their resultant surface roughness profiles measured using a Talysurf® Form Intra 50, Taylor Hobson, Leicester, UK, apparatus. Although there was a significant change in the Ra values (ranging from 0.07 to 0.26 μm) over 20 wash cycles (Figure 1), some apparent increases in roughness could not be definitively proven due to fluctuating data points. The procedure was undertaken with and without the use of caustic soda detergent but no difference in roughness was apparent. Other influences on surface roughness, such as the use of cutlery during the consumption of food, will be assessed and correlated with cleaning parameters and the detectability of residual food fouling. The implementation of this assessment process on an industrial scale will require rapid assessment methods to analyse each pack in a rapid, automated process. Future work will thus investigate the ability of rapid ultraviolet fluorescence imaging to assess the surface properties, alongside the cleaning effectiveness of washed rPET packaging, which is a technique already under development by this research team. The implications of such a system within a data-rich Industry 4.0 system are discussed.

Citation: Qaiser, M.; Woolley, E.; Colwill, J.; Hatton, F.L.; Webb, D.P. Intelligent Assessment of Reusable Plastic Food Packaging for a Circular Supply Chain. *Eng. Proc.* **2023**, *40*, 12. https://doi.org/10.3390/engproc2023040012

Academic Editor: Hana Trollman

Published: 17 July 2023

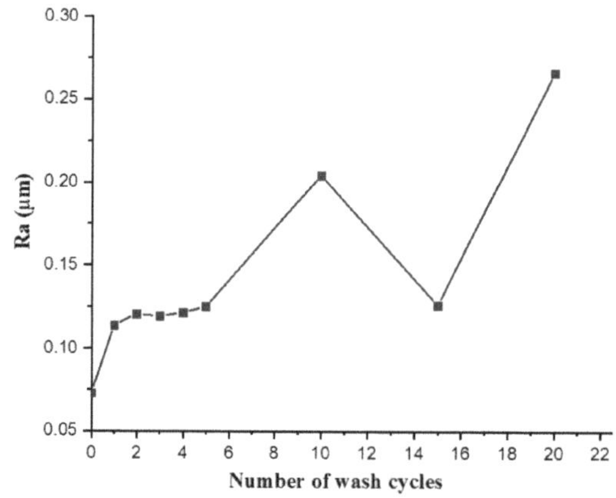

Figure 1. Roughness of rPET Packs washed without detergent at 55 °C wash and 70 °C rinse temperature.

Author Contributions: Conceptualization, E.W., F.L.H. and M.Q.; methodology, E.W. and M.Q.; formal analysis, M.Q.; investigation, M.Q.; resources, E.W.; data curation, M.Q.; writing—original draft preparation, M.Q.; writing—review and editing, E.W., J.C., D.P.W. and F.L.H.; supervision, E.W., F.L.H., J.C. and D.P.W.; project administration, E.W.; funding acquisition, E.W. All authors have read and agreed to the published version of the manuscript.

Funding: This research was partly funded by the Natural Environment Research Council, grant number [NE/V01076X/1].

Institutional Review Board Statement: Not applicable.

Informed Consent Statement: Not applicable.

Data Availability Statement: The data from this study is not currently made available.

Conflicts of Interest: The authors declare no conflict of interest.

engineering proceedings

MDPI

Abstract

The Transformation of the Agri-Food Products Export Channels from Ukraine: New Opportunities and Challenges †

Iuliia Samoilyk * and Dmytro Zhaivoron

Economy and International Economic Relations Department, Educational and Scientific Institute of Economics, Management, Low and Information Technologies, Poltava State Agrarian University, 36003 Poltava, Ukraine; dmytro.zhaivoron@pdau.edu.ua

* Correspondence: iuliia.samoilyk@gmail.com

† Presented at the International Conference on Industry 4.0 for Agri-food Supply Chains: Addressing Socio-economic and Environmental Challenges in Ukraine, Leicester, UK and Online, 24–25 July 2023.

Keywords: agri-food market; export; exports channels; supply chain; Russian war on Ukraine; "grain corridor"

The global agri-food market is one of the most important sectors of the economy, providing food for billions of people around the world. Ukraine is one of the key players in the global food market due to its high agricultural potential and large volumes of agricultural production, which several times exceed domestic needs.

The agri-food products export plays an important role in the development of the agricultural sector and the country's economy as a whole. Developed exports activities support the income of agricultural enterprises and the increase in farmers, stimulate the flow of investment, create new jobs, strengthen the country's image as a reliable quality products supplier, open access to new sales markets, and expand opportunities for development. Effective export is not possible without developed export channels, especially when exporting food products that have a limited shelf life and are picky about storage conditions. Therefore, for Ukraine and importers of its food products, under the new conditions of temporary restrictions and forced changes in export channels due to the war with Russia, their development becomes highly relevant.

Analysis of the official statistical data shows steady growth starting from March across all export channels, reaching peak values and relative stabilization from August to November 2022, with the exception of ports and railways, which continue to grow further. The sharp exports doubled in August compared to July 2022 due to the increase in the volume of exports through all channels, and the highest increase was achieved in exports through ports and sea transport, which increased by 2.1 times. Since August 2022, exports through the ports continued to grow and accounted for 60–80% of total exports, which was achieved thanks to active cooperation with the UN and Turkey in the framework of the "Grain Corridor", which allowed the operation of the ports to be unblocked and their safe use to be guaranteed. In general, the dynamics indicate a low diversification of export channels with a high dependence on ports, which makes Ukraine completely dependent on this export channel and puts it at risk as a domestic product.

Therefore, the war showed the weak points and emphasized the danger of the dependence on one export channel, causing many countries of the world, which were completely dependent on the export of Ukrainian agri-food products, to reach the brink of a food crisis. To reduce the dependence on exports through ports, Ukraine needs to pay attention to the development of other export channels, but the main and permanent importers of Ukrainian products are the countries of Asia and Africa, to which exports in a situation of war with Russia are possible only with the use of ports. To reduce the share of exports through ports, Ukraine needs to increase the volume of exports of agri-food products to

Citation: Samoilyk, I.; Zhaivoron, D. The Transformation of the Agri-Food Products Export Channels from Ukraine: New Opportunities and Challenges. *Eng. Proc.* **2023**, *40*, 13. https://doi.org/10.3390/engproc2023040013

Academic Editor: Hana Trollman

Published: 17 July 2023

EU countries and develop internal infrastructure and customs checkpoints with friendly neighboring countries. The agricultural products export from Ukraine can be related to the concept of Industry 4.0 through modern technologies and digitization implementation in the agricultural sector. Industry 4.0 encompasses the use of technologies such as the Internet of Things (IoT), artificial intelligence (AI), data analytics, automation, and robotics. These technologies can be applied in agriculture to enhance productivity, quality, and production efficiency.

For example, data analytics and artificial intelligence can be used for demand and market trends forecasting, and for optimizing the logistics of agricultural product exports. This enables proper planning of the sowing, production, and delivery of products to external markets.

Author Contributions: Conceptualization, I.S. and D.Z.; methodology, I.S. and D.Z.; software, I.S. and D.Z.; validation, I.S. and D.Z.; formal analysis, I.S. and D.Z.; investigation, I.S. and D.Z.; resources; data curation, I.S. and D.Z.; writing—original draft preparation, I.S. and D.Z.; writing—review and editing, I.S. and D.Z.; visualization, I.S. and D.Z.; supervision, I.S.; project administration, I.S.; funding acquisition, I.S. All authors have read and agreed to the published version of the manuscript.

Funding: This research was funded by the specific project "Traceability in food supply chains for improved global and national competitiveness" (https://le.ac.uk/research/projects/traceability-food-supply (accessed on 1 June 2023)).

Institutional Review Board Statement: Not applicable.

Informed Consent Statement: Not applicable.

Data Availability Statement: The data is taken from open sources and is not confidential.

Conflicts of Interest: The authors declare no conflict of interest.

engineering proceedings

MDPI

Abstract

Traceable Dairy Supply Chain Implementation in Ukraine for Improved Export Potential †

Chenyiman Lou [1,*], Roman Gorobec [2], Iuliia Samoilyk [2] and Hana Trollman [1]

1 School of Business, University of Leicester, Leicester LE2 1RQ, UK; ht203@leicester.ac.uk
2 Economy and International Economic Relations Department, Poltava State Agrarian University, 36003 Poltava, Ukraine; roman.i.horobets@pdaa.edu.ua (R.G.); iuliia.samoilyk@gmail.com (I.S.)
* Correspondence: cl521@leicester.ac.uk
† Presented at the International Conference on Industry 4.0 for Agri-food Supply Chains: Addressing Socio-economic and Environmental Challenges in Ukraine, Leicester, UK and Online, 24–25 July 2023.

Keywords: dairy supply chain; traceability; Industry 4.0 technologies; socio-economic development; food safety; food quality; Ukraine; Russia; conflict

Citation: Lou, C.; Gorobec, R.; Samoilyk, I.; Trollman, H. Traceable Dairy Supply Chain Implementation in Ukraine for Improved Export Potential. *Eng. Proc.* **2023**, *40*, 14. https://doi.org/10.3390/engproc2023040014

Academic Editor: Sandeep Jagtap

Published: 17 July 2023

This research critically examines how the traceable dairy supply chain can be implemented in Ukraine to foster socio-economic outcomes, such as reduced food safety risk, higher product quality, improved shipping and storage, and greater consumer confidence for improved export potential. The traceability of the dairy supply chain has the potential to significantly improve socio-economic conditions within countries. The Ukrainian dairy sector has experienced many fundamental and persistent problems, and such problems have been exacerbated by the Ukraine/Russia conflict. The Ukrainian dairy supply chain predominantly originates with family farm production of less than 10 cows. The Ukrainian dairy industry is primarily aimed at the domestic market, with most dairy products consumed within the country. The export of Ukrainian dairy industry products is only just developing. Concurrently, the main challenges facing the formation of supply chains for exporting Ukrainian products include securing reliable partners in other countries, forming optimal intermediary networks, and deploying Industry 4.0 technologies to ensure adequate dairy product transport and storage.

Dairy supply chains are complex systems susceptible to disruptions and unforeseen events. For the dairy sector, maintaining the safety and quality of dairy products when risks and vulnerabilities are present is critical. The 2023 dairy export forecast for Ukraine is uncertain and based on an absence of Russian advances. Despite the decreased production, the export of 2022 Ukrainian summer and fall dairy products exceeded that of 2020 and 2021. Given the highly seasonal nature, household milk is often used to produce dry milk and whey, which are then exported. Current domestic demand for dairy products remains very low due to lower disposable incomes and substantial population outflow, leading to a raw milk excess supply despite the lower production. The importance of supporting Ukraine's economy was recognized as Great Britain (from 26 April 2022) and the EU (from 4 June 2022) abolished their import duties for all Ukrainian products (including dairy) and simplified trans-shipment procedures. Available data indicate that Ukraine's export of skim milk powder (SMP) grew significantly in 2022, with butter and milk fat being the other major exports. There are 39 Ukrainian dairy production facilities approved for EU exports (with 11 new facilities accredited after the war started in 2022). Exports of butter and SMP are most attractive for Ukrainian dairy processors with high global prices driving exports despite logistics issues. Traceability promises opportunities for improved resilience and decision making throughout the supply chain, helping to provide much needed income.

This research analyzes the current state of dairy product export from Ukraine based on an interpretive sensemaking literature review to identify the socio-economic benefits of introducing traceable dairy supply chain features and processes. Attention is drawn to barriers that should be addressed in order to ensure the successful implementation of traceable dairy supply chains in Ukraine, such as the adoption of Industry 4.0 technologies, financial constraints, lack of or difficulties in enforcing regulations, lack of necessary infrastructure, and limited coordination among stakeholders. These issues warrant thoughtful consideration by policymakers and business stakeholders who wish to apply effective traceable dairy supply chain-enabled management strategies for improved export potential, especially in the context of predominantly family farm production.

Finally, the findings from this research are used to make some exploratory recommendations for the successful implementation of traceability systems in Ukraine in the near future.

Author Contributions: Conceptualization, C.L. and H.T.; methodology, H.T.; validation, C.L., R.G. and I.S.; formal analysis, H.T.; investigation, C.L.; resources, H.T.; data curation, C.L. and H.T.; writing—original draft preparation, C.L.; writing—review and editing, R.G, I.S. and H.T.; supervision, H.T.; project administration, H.T.; funding acquisition, H.T. All authors have read and agreed to the published version of the manuscript.

Funding: This research was funded by Research England and UKRI grant number 11155. The APC was funded by Research England and UKRI.

Institutional Review Board Statement: Not applicable.

Informed Consent Statement: Not applicable.

Data Availability Statement: Publicly available datasets were analyzed in this study. This data can be found here: https://www.fas.usda.gov/data/ukraine-dairy-and-products-annual-6 (accessed on 1 June 2023); https://www.fas.usda.gov/data/ukraine-dairy-and-products-annual (accessed on 1 June 2023).

Conflicts of Interest: The authors declare no conflict of interest.

engineering proceedings

Abstract

Transforming Packaging Inventory Management in the Food Industry: Unleashing the Power of Industry 4.0 for Sustainability and Resilience †

Neha Jadhav [1,*], Tom Hollands [2], Sandeep Jagtap [1], Mohamed Afy-Shararah [1] and Konstantinos Salonitis [1]

1 Sustainable Manufacturing Systems Centre, School of Aerospace Transport and Manufacturing, Cranfield University, Cranfield MK43 0AL, UK; s.z.jagtap@cranfield.ac.uk (S.J.); m.a.shararah@cranfield.ac.uk (M.A.-S.); k.salonitis@cranfield.ac.uk (K.S.)

2 Raynor Foods Ltd., 4 Farrow Road, Widford Industrial Estate, Chelmsford CM1 3TH, UK; tom.hollands@sandwiches.uk.net

* Correspondence: neha.jadhav.781@cranfield.ac.uk

† Presented at the International Conference on Industry 4.0 for Agri-food Supply Chains: Addressing Socio-economic and Environmental Challenges in Ukraine, Leicester, UK and Online, 24–25 July 2023.

Keywords: packaging material inventory management; Industry 4.0 in the food industry; sustainable practices in food packaging; supply chain resilience in the packaging sector; circular economy and waste reduction in packaging

In recent years, the need for sustainable practices within the food industry has become increasingly important and incorporating Industry 4.0 principles can further enhance these efforts. The management of packaging material inventory can greatly benefit from the integration of advanced technologies and digitalization, leading to improved sustainability and waste reduction.

This research aims to uncover the underlying causes of significant write-offs in packaging material inventory and the corresponding costs, while simultaneously exploring the potential of Industry 4.0 in addressing these challenges. By leveraging advanced technologies like Internet of Things (IoT), Artificial Intelligence (AI), and data analytics, this study can provide insights into real-time inventory monitoring, demand forecasting, and supply chain optimization.

Through an interview-based qualitative approach, the research will identify the specific areas within the packaging material life cycle where Industry 4.0 technologies can be applied. This includes exploring solutions such as smart packaging, track-and-trace systems, and predictive maintenance to mitigate factors that contribute to write-offs, such as overproduction, improper storage, and damaged inventory.

Furthermore, the research will emphasize the role of the food circular economy and how Industry 4.0 can support its implementation. By integrating circular economy principles, such as recycling, reusing, and repurposing packaging materials, this study aims to identify strategies that optimize material flows, minimize waste, and promote a more sustainable approach to inventory management.

Overall, this research not only addresses the challenges of packaging material inventory management and sustainability, but also highlights the potential of Industry 4.0 in revolutionizing these areas within the food industry. By adopting advanced technologies and embracing circular economy principles, this study aims to drive significant improvements in waste reduction, resource efficiency, and overall sustainability, ultimately contributing to a more resilient and future-ready food industry.

Citation: Jadhav, N.; Hollands, T.; Jagtap, S.; Afy-Shararah, M.; Salonitis, K. Transforming Packaging Inventory Management in the Food Industry: Unleashing the Power of Industry 4.0 for Sustainability and Resilience. *Eng. Proc.* **2023**, *40*, 16. https://doi.org/10.3390/engproc2023040016

Academic Editor: Hana Trollman

Published: 18 July 2023

Author Contributions: Conceptualization, N.J., S.J. and M.A.-S.; methodology, N.J.; validation, N.J., S.J., M.A.-S. and T.H.; formal analysis, N.J.; investigation, N.J.; resources, S.J. and T.H.; data curation, N.J. and T.H.; writing—original draft preparation, N.J., S.J., M.A.-S. and K.S.; writing—review and editing, N.J., S.J., M.A.-S. and K.S.; visualization, N.J. and T.H.; supervision, S.J., M.A.-S. and K.S.; project administration, S.J., M.A.-S., K.S. and T.H.; funding acquisition, K.S. All authors have read and agreed to the published version of the manuscript.

Funding: This research received no external funding.

Institutional Review Board Statement: Not applicable.

Informed Consent Statement: Not applicable.

Data Availability Statement: Not applicable.

Conflicts of Interest: The authors declare no conflict of interest.

engineering proceedings

MDPI

Abstract

The Role of Ukrainian Innovation Centres in the Development of the Agricultural Sector and Supply Chains †

Nataliia Demianenko [1,*], Olena Krasota [2] and Ilona Yasnolob [1]

1 Department of Entrepreneurship and Law, Educational and Scientific Institute of Economics, Management, Law and Information Technologies, Poltava State Agrarian University, 36003 Poltava, Ukraine; ilona.yasnolob@pdaa.edu.ua
2 Faculty of Accounting and Finance, Department of Humanities and Social Sciences, Poltava State Agrarian University, 36003 Poltava, Ukraine; olena.krasota@pdaa.edu.ua
* Correspondence: nataliia.bredun@pdaa.edu.ua

† Presented at the International Conference on Industry 4.0 for Agri-food Supply Chains: Addressing Socio-economic and Environmental Challenges in Ukraine, Leicester, UK and Online, 24–25 July 2023.

Keywords: Supply Chain 4.0; Industry 4.0; start-ups; agricultural sector; Center's development; modelling and optimisation of business processes

The global challenges of 2019–2023 have impacted supply chains. Global supply chains, disrupted by Russian aggression, affect countries not only in Africa but around the world. Many agricultural enterprises in Ukraine have been damaged. Therefore, Ukraine's exports have declined. Agro-food products remain the most exported goods among the sectors, representing 53% of total exports in 2022. Also, 2022 has revealed the benefits of domestic processing: companies aimed at exporting raw materials had to decrease their activities or stop them altogether due to the blockade of ports on the Black Sea and limited capacities of Ukrzaliznytsia. The availability of well-functioning logistics, access to railways and the company's closeness to western borders meant efficient production for Ukrainian companies in 2022. It is worth betting on start-ups to emerge from the crisis in Industry 4.0. The Global Industry 4.0 will generate up to $ 1 trillion and most technological innovation will come from innovative new start-ups by 2025.

Start-ups are identified as drivers of innovation in Industry 4.0, especially in emerging technologies. Start-ups can develop new ideas and technologies faster and more precisely in emerging technology areas. In particular, start-up developers from Ukraine presented their products to entrepreneurs and investors at the Startup Grind Global 2023 conference in Silicon Valley. The scientists presented INPUT SOFT (InputSoft Inc., Kyiv, Ukraine), a platform for managing airports and airlines; FuelWell (FUELWELL, Wilmington, DE, USA), equipment that enables trucks to consume up to 20% less fuel and emit 15% less CO_2; Adminix (Adminix Solutions Inc., Kharkiv, Ukraine), a low-code platform for work-flow automation; efarm.pro (Gardarika Tres LLC, Kyiv, Ukraine), an IOT navigation field assistant; Knowledgator Engineering (Knowledgator Engineering LTD, London, UK), a market, scientific and competitive analysis tool to improve R&D investment, etc. Conditions, environment and development vector are necessary for the rapid growth of startups. For this purpose, the Industry 4.0 Implementation Center was established at Poltava State Agrarian University. The Center is the regional core of Industry 4.0 competencies and one of the elements of the innovation ecosystem of the agricultural sector in the Poltava region of Ukraine. It is a scientific hub that explores topical issues of smart specialisation in agro-industry and supply chain logistics. In the scientific centre, research is carried out by teachers and students, and cooperations with entrepreneurs and communities are established. Start-ups are being developed to solve identified problems. Teachers provide training in innovation activities for representatives of territorial communities, and small-

Citation: Demianenko, N.; Krasota, O.; Yasnolob, I. The Role of Ukrainian Innovation Centres in the Development of the Agricultural Sector and Supply Chains. *Eng. Proc.* **2023**, *40*, 17. https://doi.org/10.3390/engproc2023040017

Academic Editor: Hana Trollman

Published: 18 July 2023

and medium-sized enterprises. The institution operates a laboratory where research and testing of innovative products is carried out. In particular, a development project was implemented to automate grain receiving points. A significant number of businesses from different regions of Ukraine have relocated to the Poltava region. This emphasises the importance of the Center's development.

According to Supply Chain 4.0, supply chain management is based on "Industry 4.0" innovations, namely the internet of things, robotics, analytics and big data, which aim to increase productivity and customer satisfaction. In the near future, Supply Chain 4.0 will obviously affect all areas of supply chain management. The development prospects of the Poltava region of Ukraine in the context of digitalisation of the agro-industry envisage the accelerated digitalisation of agricultural and agro-processing enterprises and export logistics. This is very important in the Poltava region, which produces the second-largest soybean and sunflower harvests in Ukraine. For this purpose, it is necessary to enhance the role of clusters and develop regional Industry 4.0 implementation centres. The University Center will take part in joint training on the implementation of industrial recovery and innovation development projects in Ukraine. The main trends in the formation of Ukrainian supply chains that will also be addressed by the Center in 2023–2024 should be the modelling and optimisation of business processes, digital transformation of supply chains, big data and management of digital platforms, IT start-ups and the future of intelligent logistics, and the further transition to "green" solutions. Ukraine requires integration into the global digital ecosystem.

Author Contributions: N.D.: writing—original draft preparation; I.Y.: validation; O.K.: validation and translation. All authors have read and agreed to the published version of the manuscript.

Funding: This research received no external funding.

Institutional Review Board Statement: Not applicable.

Informed Consent Statement: Informed consent was obtained from all subjects involved in the study.

Data Availability Statement: Industry4-0-ukraine URL: https://industry4-0-ukraine.com.ua/2018/09/17/stavka-na-startapi-v-%D1%96ndustr%D1%96%D1%97-4-0-v-chomu/ (accessed on 10 May 2023); pdauindustry4 URL: https://pdauindustry4.pdau.edu.ua/ (accessed on 20 May 2023); Startup Grind Global 2023 URL: https://speka.media/stalo-vidomo-yaki-ukrayinski-startapi-poyidut-na-startup-grind-global-2023-u-kremnijevu-dolinu-vz006p (accessed on 1 May 2023); Agricultural products are the most exported good URL: https://agrotimes.ua/agromarket/agroprodukcziya-ye-najbilsh-eksportovanym-tovarom/ (accessed on 21 May 2023).

Conflicts of Interest: The authors declare no conflict of interest.

engineering proceedings

MDPI

Abstract

Addressing Unfairness in Fresh Fruit Supply Chains in the United Kingdom with Technology Adoption for Improved Supply Chain Resilience †

Adenike A. Moradeyo *, Adegboyega Oyedijo and Hana Trollman

School of Business, University of Leicester, Leicester LE2 1RQ, UK; a.oyedijo@le.ac.uk (A.O.); ht203@leicester.ac.uk (H.T.)

* Correspondence: aam77@leicester.ac.uk

† Presented at the International Conference on Industry 4.0 for Agri-food Supply Chains: Addressing So-cio-economic and Environmental Challenges in Ukraine, Leicester, UK and Online, 24–25 July 2023.

Keywords: big data; blockchain; supply chain resilience; fairness; food industry; creativity; supply chain disruption

This research aims to develop a better understanding of how the adoption of Industry 4.0 technologies, specifically the use of big data (BD) and blockchain, can help from the standpoint of creativity to promote responsible and ethical procurement and supply chain practices. In particular, we aim to explain how creativity can be employed as an alternative supply chain resilience strategy to address issues of unfairness to support the resilience of fresh fruit supply chains in United Kingdom (UK). Overall, the aim is to minimize potential of disruptions. In the present era, disruptive events and ongoing challenges have considerably hindered the ability of businesses to produce and distribute their goods. Such events, to mention only several, include COVID-19, terrorism, natural disasters, political unrest, and economic issues with potential national and global implications. Moreover, the aforementioned disruptions have heightened the cognizance of scholars and professionals alike regarding the need for greater resilience in supply chains as a means of alleviating potentially catastrophic consequences. Much emphasis has been placed on addressing the negative impacts of tangible, visible, and catastrophic events through building more resilient supply chains. However, intangible but disruptive and prevalent occurrences such as unfairness in supply chains, although being addressed, are still in the infancy phase of being resolved with respect to supply chain resilience. Furthermore, such intangible disruptions tend to be aggravated by tangible events. In particular, there is a paucity of prior research on the connection between supply chain unfairness and resilience. The presence of unfair practices within the supply chains of the agri-food industry has the potential to generate food insecurity. In this research, unfairness is addressed as a disruptive intangible event affecting supply chains that requires management. It has been posited that risk categorization is key to identifying the relevant mitigation strategies. Specifically, in this study, unfairness to suppliers in the agri-food supply chain is a form of demand risk because it emanates from the downstream part of the supply chain.

The study will investigate how industry 4.0 technologies' adoption, specifically, big data and blockchain, can mitigate unfairness and support creativity in agri-food supply chains as a practical application of alternative strategies for cultivating supply chain resilience beyond the established traditional strategies or capabilities of flexibility, redundancy, collaboration, and agility. The specific case considered within the context of the UK is fresh fruit supply chains as affected during Brexit. According to the constraint–creativity paradigm, it is possible to develop creativity from constraints. Therefore, in this study, the constraint–creativity paradigm informs the practical implementation of BD and blockchain technologies to promote the creativity that enhances supply chain resilience. Creativity as a

Citation: Moradeyo, A.A.; Oyedijo, A.; Trollman, H. Addressing Unfairness in Fresh Fruit Supply Chains in the United Kingdom with Technology Adoption for Improved Supply Chain Resilience. *Eng. Proc.* **2023**, *40*, 18. https://doi.org/10.3390/engproc2023040018

Academic Editor: Iuliia Samoilyk

Published: 18 July 2023

strategy to enhance supply chain resilience is based on the tenets of the extended resource base view (RBV) theory. Creativity that emanates from industry 4.0 technologies' adoption is treated as an intangible resource or capability in this study. This research employs the extended RBV theory to address the issue of supply chain resilience using a mix of methods. This is achieved by leveraging the tangible and intangible benefits of industry 4.0 technologies' adoption to effectively manage supply chain disruptions (unfairness in this case), thereby enhancing the overall resilience of the supply chain.

Author Contributions: Conceptualization, A.A.M., A.O. and H.T.; writing—A.A.M.; writing—review and editing, A.A.M., A.O. and H.T.; supervision A.O. and H.T.; funding acquisition, H.T. All authors have read and agreed to the published version of the manuscript.

Funding: The APC was funded by Research England and UKRI.

Institutional Review Board Statement: Not Applicable.

Informed Consent Statement: Not Applicable.

Data Availability Statement: No new data were created or analyzed in this study. Data sharing is not applicable to this article.

Conflicts of Interest: The authors declare no conflict of interest

engineering proceedings

Abstract

Resilient Food Supply Chains in the Face of the Russo–Ukrainian War: Harnessing the Power of the Internet of Things †

Guillermo Garcia-Garcia [1,*], Carlos Parra-López [1], Sandeep Jagtap [2] and Hana Trollman [3]

1 Department of Agrifood System Economics, Institute of Agricultural and Fisheries Research & Training (IFAPA), 18080 Granada, Spain; carlos.parra@juntadeandalucia.es
2 Sustainable Manufacturing Systems Centre, School of Aerospace, Transport and Manufacturing, Cranfield University, Cranfield MK43 0AL, UK; s.z.jagtap@cranfield.ac.uk
3 School of Business, University of Leicester, Leicester LE2 1RQ, UK; ht203@leicester.ac.uk
* Correspondence: guillermo.garcia@juntadeandalucia.es
† Presented at the International Conference on Industry 4.0 for Agri-food Supply Chains: Addressing Socio-economic and Environmental Challenges in Ukraine, Leicester, UK and Online, 24–25 July 2023.

Abstract: The current war in Ukraine has severely disrupted global food supply chains due to the significant decline in the production of grain commodities, of which Ukraine and Russia are major global suppliers. This creates problems in today's globalised food systems. In particular, a number of countries are heavily dependent on food imports from Ukraine and/or Russia, particularly some developing nations in regions like the Middle East and North Africa. As a result, alternative suppliers need to increase production and food supply chains need to adapt distribution processes. In these situations, building resilience in food supply chains is critical to react to disruptions. Resilience has become a widely researched topic in the context of food supply chains, leading to the exploration of different strategies. One potential strategy to improve resilience is to integrate digital technologies to optimise food operations. One of the digital technologies at the core of Industry 4.0, and one that is increasingly being used, is the Internet of Things (IoT). The IoT refers to systems that connect devices via the Internet, allowing them to collect and share data in real time within a network. The successful use of IoT in industrial systems has already been demonstrated and the food industry has begun to embrace the opportunities this technology offers. This research focuses on the potential of the IoT to support food supply chains by making them more resilient, particularly in the context of the disruptions caused by the war in Ukraine. Using a literature review and data from statistical sources, we explore the challenges posed to food supply chains by the Russo–Ukrainian war, and identify specific vulnerabilities that the IoT can address. We discuss the challenges of implementing this technology, particularly in wartime situations, and its potential to strengthen food supply chain operations through its ability to collect and share accurate data in real time.

Keywords: Russo–Ukrainian war; food supply chains; food production disruption; resilience; Internet of Things (IoT)

Citation: Garcia-Garcia, G.; Parra-López, C.; Jagtap, S.; Trollman, H. Resilient Food Supply Chains in the Face of the Russo–Ukrainian War: Harnessing the Power of the Internet of Things. *Eng. Proc.* **2023**, *40*, 19. https://doi.org/10.3390/engproc2023040019

Academic Editor: Iuliia Samoilyk

Published: 20 July 2023

Author Contributions: Conceptualization, G.G.-G.; methodology, G.G.-G.; formal analysis, G.G.-G.; investigation, G.G.-G.; writing—original draft preparation, G.G.-G.; writing—review and editing, G.G.-G., C.P.-L., S.J. and H.T.; supervision, C.P.-L. All authors have read and agreed to the published version of the manuscript.

Funding: This research received no external funding.

Institutional Review Board Statement: Not applicable.

Informed Consent Statement: Not applicable.

Data Availability Statement: No new data were created or analyzed in this study. Data sharing is not applicable to this article.

Conflicts of Interest: The authors declare no conflict of interest.

engineering proceedings

MDPI

Abstract

Features and Prospects of Industry 4.0 in the Agrarian Sector of Ukraine in Wartime: Economic and Accounting Aspects †

Liudmyla Chip [1,*], Natalia Kantsedal [1], Olena Kopishynska [2], Yuri Utkin [2] and Olena Taran-Lala [2]

1 Faculty of Accounting and Finance, Poltava State Agrarian University, 36003 Poltava, Ukraine; nataliia.kantsedal@pdaa.edu.ua
2 Educational and Scientific Institute of Economics, Management, Law and Information Technologies, Poltava State Agrarian University, 36003 Poltava, Ukraine; olena.kopishynska@pdaa.edu.ua (O.K.); utkin@pdaa.edu.ua (Y.U.); olena.taran-lala@pdaa.edu.ua (O.T.-L.)
* Correspondence: liudmyla.chip@pdaa.edu.ua
† Presented at the International Conference on Industry 4.0 for Agri-food Supply Chains: Addressing Socio-economic and Environmental Challenges in Ukraine, Leicester, UK and Online, 24–25 July 2023.

Keywords: Industry 4.0; agricultural technology; food security; agri-food supply chains; databases; accounting

Citation: Chip, L.; Kantsedal, N.; Kopishynska, O.; Utkin, Y.; Taran-Lala, O. Features and Prospects of Industry 4.0 in the Agrarian Sector of Ukraine in Wartime: Economic and Accounting Aspects. *Eng. Proc.* **2023**, *40*, 20. https://doi.org/10.3390/engproc2023040020

Academic Editor: Hana Trollman

Published: 20 July 2023

The agrarian sector of Ukraine is an integral part of the country's economy. As a result of the war with the Russian Federation, the country's economic situation is characterized by a decrease in the production output, shutdowns of some industries, and a significant drop in consumption. The main challenges faced by the enterprises are the gaps in logistics supplies, loss of personnel, falling effective demand, loss of significant areas of agricultural land due to temporary occupation, decrease in soil fertility as a result of hostilities, lack of necessary capital and the difficulties in predicting the situation in general. These factors negatively affected the food supply chains and led to their disruption. In wartime, the solution to the problem of the country's food security is extremely important. Maintaining the sustainability of agribusiness and its competitiveness is a priority task for the state in ensuring food security, supported by the innovative technologies of the Industry 4.0 concept [1,2].

At the present time, there is an urgent task to determine and assess the consequences and possibilities of further development of agribusiness at the local and global levels using the innovative nature of agricultural production and the economy on the whole. The use of modern systems of automation of agricultural production management, digitalization of large volumes of data, and use of other modern technologies will allow to increase the efficiency of production and planning of economic growth in territories that have not experienced the destruction of the system [3]. Obtaining objective and operational data will make it possible to apply AI algorithms to make an informed decision in the management of technological processes in the agricultural sector. At the same time, the introduction of innovative technologies will allow to return the level of technological production of agricultural products to the pre-war period in the de-occupied territories [4]. Such development is based on the principles of caring for the environment and at the same time achieving the maximum economic effect from agricultural activity [5]. An important condition for attaining food security is the stability of its provision, which gives the possibility for groups of the population of households and individuals to have access to sufficient amounts of food at any time and not be under the pressure of losing access to food as a result of supply and demand shocks.

Thus, Industry 4.0 should become the state priority, which determines whether the country will be able to demonstrate the sustainable development of agribusiness and achieve high rates of social and economic development. Therefore, in order to remain competitive in domestic and foreign markets, even in wartime, agribusiness approves solutions in the context of the development of Industry 4.0. The approach, which illustrates the synergy of Agriculture 4.0 and Economy 4.0 technologies, will ensure the growth of economic, environmental, and social efficiency as a progressive direction of the local and global economic system, can be applied in wartime. In conditions of uncertainty and growing threats, the general model of a new informational paradigm for the activity of agricultural organizations, which developed on the basis of knowledge mobility and the transfer of analytical data into valuable information, is of particular importance. The methodology for evaluating the interaction of information agents in agricultural production is based on the model of previous studies [5].

The prospect of further research involves a further in-depth and comprehensive analysis of the ability of the information system to accumulate and interpret data on activities in agricultural production, determination of chains of interaction, methods of interpretation, and analysis of information flows at both local and global levels.

Author Contributions: Conceptualization, L.C., N.K., O.K., Y.U. and O.T.-L.; methodology, L.C., N.K., O.K., Y.U. and O.T.-L.; software, O.K., Y.U., L.C., N.K. and O.T.-L.; validation, L.C., N.K., O.K., Y.U. and O.T.-L.; formal analysis, O.K., Y.U., L.C. and N.K.; investigation, L.C., N.K., O.K., Y.U. and O.T.-L.; resources, L.C., N.K., O.K., Y.U. and O.T.-L.; data curation, L.C., N.K., O.K. and Y.U.; writing—original draft preparation, L.C., N.K., O.K. and Y.U.; writing—review and editing, L.C.; supervision, L.C., N.K., O.K. and Y.U.; project administration, L.C. All authors have read and agreed to the published version of the manuscript.

Funding: This research received no external funding.

Institutional Review Board Statement: Not applicable.

Informed Consent Statement: Not applicable.

Data Availability Statement: Data on the results of the real state of the agricultural sector of Ukraine are taken from official websites: Ministry of Agrarian Policy and Food of Ukraine https://minagro.gov.ua/ (accessed on 1 June 2023), State Statistics Service of Ukraine https://ukrstat.gov.ua/ (accessed on 1 June 2023), UCAB https://www.ucab.ua/en/ (accessed on 1 June 2023). The work contains references to the previous studies of the authors, which are the basis for continuation in this research. The composition of the frameworks of Industry 4.0 technologies and their characteristics are publicly available, in particular in scientific publications.

Conflicts of Interest: The authors declare no conflict of interest.

References

1. Kharchuk, T.; Tarasenko, I.; Chip, L.; Sakun, L.; Bebko, S.; Palchuk, O.; Yatsykovskyy, B.; Ishejkin, T.J. Modeling the concept of managing changes of the smart economy and financial stability of entrepreneurship under martial law. *Hyg. Eng. Des.* **2022**, *40*, 239–249.
2. Kibik, O.; Taran-Lala, O.; Saienko, V.; Metil, T.; Umanets, T.; Maksymchuk, I. Strategic Vectors for Enterprise Development in the Contex of the Digitalization of the Economy. *Postmod. Openings* **2022**, *13*, 384–395. [CrossRef]
3. Kopishynska, O.; Utkin, Y.; Lyashenko, V.; Barabolia, O.; Kalashnik, O.; Mororz, S.; Kartashova, O. Information System and Technologies in Agronomy and Business: Employers' Requirements-Oriented Study in Agricaltural Universities. In Proceedings of the 25th World Multi-Conference on Systemics, Cybernetics and Informatics (WMSCI 2021), Online, 18–21 July 2021; pp. 113–118.
4. Kopishynska, O.; Utkin, Y.; Galych, O.; Marenych, M.; Sliusar, I. Main Aspects of the Creation of Managing Information System at the Implementation of Precision Farming. In Proceedings of the IEEE 11th International Conference on Dependable Systems, Services and Technologies (DESSERT), Kyiv, Ukraine, 14–18 May 2020; pp. 404–410. [CrossRef]
5. Kantsedal, N.; Ponomarenko, O. Institutional Model of Integrating Agricultural Production Technologies with Accounting and Informational Systems. In *The Digital Agricultural Revolution:Innovations and Challenges in Agriculture through Technology Disruptions*; Wiley-Scrivener: Beverly, MA, USA, 2022; pp. 301–310. [CrossRef]

engineering proceedings

MDPI

Abstract

Leveraging Industry 4.0 for Supply Chain Collaboration: Creating Competitive Advantage for Small Farms in the United Kingdom and Ukraine †

Leila Masoomi [1], Olena Taran-Lala [2], Charlotte V. L. Smith [1] and Hana Trollman [1,*]

1 School of Business, University of Leicester, Leicester LE2 1RQ, UK; lm557@leicester.ac.uk (L.M.); cvls1@leicester.ac.uk (C.V.L.S.)
2 Entrepreneurship and Law Department, Poltava State Agrarian University, 36003 Poltava, Ukraine; elenalala1704@gmail.com
* Correspondence: ht203@leicester.ac.uk

† Presented at the International Conference on Industry 4.0 for Agri-food Supply Chains: Addressing Socio-economic and Environmental Challenges in Ukraine, Leicester, UK and Online, 24–25 July 2023.

Keywords: Industry 4.0; supply chain collaboration; competitive advantage; small farms; resource-based view; social network theory

The aim of this research is to comparatively analyse the potential for implementing Industry 4.0 technologies in agri-food supply chains for the benefit of small farms in the United Kingdom (UK) and Ukraine with a focus on enhancing their capabilities and competitiveness. The promotion of local food systems and short supply chains is supported for enhancing the resilience of the food system, particularly during crises, while also improving environmental impact. Local farms play a crucial role in minimizing the environmental consequences associated with long-distance food transportation. Additionally, they positively impact the local economy by supporting farmers and businesses, and foster sustainable agriculture practices through supply chain transparency. However, small- and medium-sized enterprises (SMEs) and small farms face specific constraints compared to larger enterprises. These constraints include reduced economies of scale, increased costs, limited resources, and low efficiency. Poor linkages and communication in the agri-food supply chain exacerbate these challenges, whilst cyber security and integration with legacy systems may not be compatible with new Industry 4.0 technologies. Despite the necessity to adapt to Industry 4.0 and utilize digital technologies in agriculture, obstacles persist, including limited technological literacy, inadequate human resources, network coverage, and capital support, which hinder the full potential of smart agriculture in both the UK and Ukraine.

Considering the specific needs and constraints of each farm when planning and implementing Industry 4.0 technologies is critical. In addition, government support, education and access to finance can contribute to the successful implementation of Industry 4.0 in British and Ukrainian agriculture. According to the resource-based view (RBV), organizations have the capacity to develop dynamic capabilities by utilizing and reconfiguring their resources. Social network theory complements the RBV, emphasizing the role of network structure and partner characteristics in gaining access to external resources. This study adopts a comparative analysis of small farms in the UK and Ukraine based on the application of the RBV and social network theory. An analysis of how small farms in the UK and Ukraine can create competitive advantage by adopting blockchain and IoT technologies in vertical and horizontal supply chain collaboration is presented. By leveraging technology and enhancing collaboration, small farms in the UK and Ukraine can overcome resource constraints, improve efficiency, and achieve sustainable growth in a competitive market environment.

Citation: Masoomi, L.; Taran-Lala, O.; Smith, C.V.L.; Trollman, H. Leveraging Industry 4.0 for Supply Chain Collaboration: Creating Competitive Advantage for Small Farms in the United Kingdom and Ukraine. *Eng. Proc.* **2023**, *40*, 21. https://doi.org/10.3390/engproc2023040021

Academic Editor: Iuliia Samoilyk

Published: 20 July 2023

Author Contributions: Conceptualization, L.M., O.T.-L. and H.T.; methodology, L.M. and H.T.; writing—original draft preparation, L.M.; writing—review and editing, O.T.-L. and H.T.; supervision, C.V.L.S. and H.T.; funding acquisition, H.T. All authors have read and agreed to the published version of the manuscript.

Funding: The APC was funded by Research England and UKRI.

Institutional Review Board Statement: Not applicable.

Informed Consent Statement: Not applicable.

Data Availability Statement: No new data were created or analysed in this study. Data sharing is not applicable to this article.

Conflicts of Interest: The authors declare no conflict of interest.

engineering proceedings

Abstract

Innovative Projects in the Industry 4.0 Sphere of Poltava State Agrarian University †

Volodymyr Muravlov *, Yurii Utkin, Ihor Sliusar, Olena Kopishynska, Oleksii Goryk, Oleksandr Brykun, Anatolii Semenov, Serhii Bida, Oleksandr Petrash and Ruslan Petrash

Industry 4.0 Implementation Center, Poltava State Agrarian University, Skovorody Str. 1/3, 36003 Poltava, Ukraine; utkin@pdau.edu.ua (Y.U.); ihor.sliusar@pdau.edu.ua (I.S.); olena.kopishynska@pdau.edu.ua (O.K.); oleksii.goruk@pdau.edu.ua (O.G.); oleksandr.brykun@pdau.edu.ua (O.B.); anatolii.semenov@pdau.edu.ua (A.S.); serhii.bida@pdau.edu.ua (S.B.); oleksandr.petrash@pdau.edu.ua (O.P.); ruslan.petrash@pdau.edu.ua (R.P.)

* Correspondence: volodymyr.muravlov@pdau.edu.ua

† Presented at the International Conference on Industry 4.0 for Agri-food Supply Chains: Addressing Socio-economic and Environmental Challenges in Ukraine, Leicester, UK and Online, 24–25 July 2023.

Keywords: industry 4.0; neural networks; electrotechnical systems; air disinfection; robotic complex; monitoring groundwater

The activity of the "Industry 4.0 Implementation Center at PSAU" is aimed at the development and implementation of promising technologies, appropriate specialists training, implementation and management of informational and educational events, training management, and networking of enterprises: technology designers, production facilities, higher-education institutions, scientific institutions, state authorities, and civil groups. The latest projects that are being implemented are as follows:

Unified information platform for the management of material resources of territorial communities. This project is aimed at creating a unified digital information space using a cloud ERP for managing all processes and resources of territorial communities as non-industrial enterprises in the context of building a landscape of Industry 4.0 technologies. Domestic ERP Universal offers the advantage of being able to manage enterprises with different accounting schemes in one database, makes it possible to scale the system to most enterprises in the future, and uses a combination of artificial intelligence and Internet of Things technologies.

Segmentation of analog meter readings using neural networks. One of the constraints to the implementation of the concepts of Smart City, Industry 4.0, etc., is the need to integrate analog energy metering. Their replacement with digital devices is often not cost-effective. One option to overcome this barrier could be a combination of AI and IoT technologies. The project proposes options for solving the image segmentation problem of displaying digital indicators of analogue meters using neural networks.

Electrotechnical systems for bactericidal air disinfection. Air often becomes a source and transmitter of pathogenic microorganisms, infections, and viruses. Ultraviolet radiation in the range of 200 to 280 nm, depending on the dosage of exposure, inactivates all microorganisms. The development of electrotechnical systems for UV disinfection (ozone-free) will provide a comprehensive solution to the task of bactericidal air disinfection in various spheres of human activity.

Complex of shot blasting processing of metal products. In mechanical engineering, there is a tendency to replace expensive metals in the manufacture of large-sized products with more affordable carbon steels with corrosion-resistant non-metallic coatings on working surfaces. The strength of such coatings is ensured by the quality of the surface preparation achieved by shot blasting. A robotic complex of remote processing of metal surfaces of free forms with compliance with the specified parameters is offered.

Citation: Muravlov, V.; Utkin, Y.; Sliusar, I.; Kopishynska, O.; Goryk, O.; Brykun, O.; Semenov, A.; Bida, S.; Petrash, O.; Petrash, R. Innovative Projects in the Industry 4.0 Sphere of Poltava State Agrarian University. *Eng. Proc.* **2023**, *40*, 22. https://doi.org/10.3390/engproc2023040022

Academic Editor: Hana Trollman

Published: 24 July 2023

Groundwater level monitoring automation. To control groundwater flooding, it is necessary to determine its cause system. This can be done by analyzing the groundwater table depth change. Conducting such research is a long and time-consuming process, so it is necessary to create an automated system for monitoring groundwater. The placement of sensors that will automatically provide data on the groundwater level rise or fall in large areas over periods of time will enable the identification of flooding causes and the development of effective control measures.

Author Contributions: Conceptualization, V.M., O.K. and Y.U.; methodology, O.K., O.G. and A.S.; software, Y.U., I.S. and O.K.; validation, O.K., I.S., S.B. and O.B.; formal analysis, O.K., Y.U., O.G. and A.S.; investigation, O.K., I.S., A.S., O.B. and S.B.; resources, O.K., Y.U., V.M.; data curation, O.K., A.S., O.G.; writing—original draft preparation, O.P., R.P., V.M.; writing—review and editing, V.M. and O.P.; supervision, V.M. and Y.U.; project administration, V.M. All authors have read and agreed to the published version of the manuscript.

Funding: This research received no external funding.

Institutional Review Board Statement: Not applicable.

Informed Consent Statement: Not applicable.

Data Availability Statement: Information on presented projects of the Industry 4.0 Implementation Center is also publicly available on the website https://pdauindustry4.pdau.edu.ua/en/home-eng/ (accessed on 1 June 2023), including in scientific publications.

Conflicts of Interest: The authors declare no conflict of interest.

engineering proceedings

Abstract

Industry 4.0 Challenges Facing the Agri-Supply Chain: A Literature Review †

Walid Khalid Al Saad *, Moayad Al-Talib and Simon Peter Nadeem

Center for Supply Chain Improvement, University of Derby, Derby DE22 1GB, UK; modtalib87@gmail.com (M.A.-T.); s.nadeem@derby.ac.uk (S.P.N.)
* Correspondence: w.alsaad@derby.ac.uk
† Presented at the International Conference on Industry 4.0 for Agri-food Supply Chains: Addressing Socio-economic and Environmental Challenges in Ukraine, Leicester, UK and Online, 24–25 July 2023.

Keywords: Industry 4.0; innovation; sustainability; agri-supply chain

The world is moving fast, and organizations are facing new risks, uncertainties, and threats within their supply chains. Food security, waste minimization, and sustainability are new strategic requirements agri-organizations must address and focus on. Nevertheless, due to globalization and changes in the market post-COVID-19, supply chains are becoming more complex and agri-organizations must change their business processes and operations to become more competitive, responsive, and agile.

To be more customer-oriented, efficient, and sustainable, agri-organizations have been adopting new innovative technologies such as RFID and the Internet of Things. Recently, the concept of Industry 4.0 (I4.0) has drawn significant attention from specialists, academics, and decision-makers. I4.0 technologies can enhance supply chain performance by bringing new cost-effective, green lean features and solutions that can improve internal and external event traceability, process automation, waste management, data sharing, and KPI monitoring.

However, there are many challenges to the adoption of I4.0 technologies. Therefore, it is very important to study these challenges facing the agri-supply chain.

The main aim of this research is to analyze the literature through a systematic approach to highlight and identify the challenges agri-supply chains are facing while deciding on implementing I4.0 technologies. Using a set of 64 publications, including journal papers, conference papers, reviews, and books chapters, selected from the well-known database Scopus for the last 7 years, a systematic literature review was conducted. The SLR highlights and considers several I4.0 technologies and their applications affecting the agri-supply chain.

The results of this study are based on a thorough analysis of the added-value outcomes of implementing different I4.0 technologies in the agri-food supply chain. Over a dozen challenges are remarked upon in the literature. These challenges are categorized into three main areas in the agri-food supply chain: infrastructural, technical and operational. The most critical challenges detected—within most observations—are technological architecture, security and privacy, and internet- and IoT-based infrastructure.

Citation: Al Saad, W.K.; Al-Talib, M.; Nadeem, S.P. Industry 4.0 Challenges Facing the Agri-Supply Chain: A Literature Review. *Eng. Proc.* **2023**, *40*, 25. https://doi.org/10.3390/engproc2023040025

Academic Editor: Hana Trollman

Published: 2 August 2023

Author Contributions: Conceptualization, W.K.A.S.; methodology, and formal analysis, W.K.A.S.; data curation and writing—original draft preparation, W.K.A.S.; writing—review and editing, W.K.A.S. and M.A.-T.; supervision, S.P.N. All authors have read and agreed to the published version of the manuscript.

Funding: This research received no external funding.

Institutional Review Board Statement: Not applicable.

Informed Consent Statement: Not applicable.

Data Availability Statement: No new data are created.

Conflicts of Interest: The authors declare no conflict of interest.

engineering proceedings

Abstract

Transformation of the Ukrainian Agri-Food Industry in the Context of Global Digitalization †

Svitlana Tul [1,*], Iuliia Samoilyk [2], Vita Klymenko [1] and Olha Shkurupii [2]

1 Regional Development Agency of Poltava Oblast "Office of European Integration", 36000 Poltava, Ukraine; vita.klimenko.85@gmail.com
2 Economics and International Economic Relations Department, Poltava State Agrarian University, 36000 Poltava, Ukraine; iuliia.samoilyk@gmail.com (I.S.); olga.sh0123@gmail.com (O.S.)
* Correspondence: tulsvetlana5@gmail.com
† Presented at the International Conference on Industry 4.0 for Agri-food Supply Chains: Addressing Socio-economic and Environmental Challenges in Ukraine, Leicester, UK and Online, 24–25 July 2023.

Abstract: Nowadays, the agri-food sector is facing fundamental challenges. According to the FAO study, the amount of arable land per capita in the world will decrease from 0.6 hectares per person in 2000 to 0.2 hectares by 2050, while the demand for food will increase by 70%. With today's yield growth of 1.5% per year, such changes could result in global food shortages. Therefore, the governments of developed and developing countries should support initiatives for the digitization of agri-food businesses and the introduction of new technologies to increase the volume of food production. Russia's war against Ukraine is the main cause of the global food crisis, which could bring serious political and economic consequences. The agricultural and food sector of Ukraine is about 10% of GDP. For many years, the Ukrainian agro-industrial complex, before the full-scale invasion of Russia, occupied a leading position among the global exporters. Ukraine supplied 10% of world wheat exports, more than 14% of corn and more than 47% of sunflower oil. A full-scale war has become a real test for the Ukrainian agri-food industry. The invasion entailed the destruction of food production processes and logistics chains. Many sowing areas were mined, equipment and warehouses were destroyed. At the end of 2022, Ukraine exported agricultural products worth USD 23.6 billion. Although the figure for 2022 is 15% less than the record of 2021 (USD 27.9 billion), last year's value of exports became the second since the independence of Ukraine. Disruptions to Ukrainian exports exacerbated the rise in food prices, which, according to the FAO index, increased by 54% in February 2022. In March 2023, prices fell, but they were still 6.4% higher than in 2022. The purpose of the study is to assess the level of digital transformation of the Ukrainian agri-food industry in order to ensure food security at the national and international levels. Digitalization of the agri-food industry in Ukraine should be considered a source of deep systemic transformations, which involves the use of digital technologies at the business level to optimize business operations, increase company productivity, and improve interaction with suppliers and customers. For agri-food companies, the issue of digitalization concerns not only technological modernization, but also a complete change of business processes: farm management systems, data processing and harvest forecasting, agricultural processing, food quality management, systems for creating added value for products, warehouse management systems, and human resources management. Nowadays, digitalization can accelerate the transformation of the agri-food industry across the entire supply chain, from manufacturing and purchasing processes to distribution, logistics and finance. Innovative technologies that can become breakthrough in the agri-food industry are as follows: bioinformatics; synthetic biology; food design; smart farming; vertical farms; aquaculture; bioinformatics; genetics; alternative sources of protein; technology of conservation and extension of the shelf life of food products. In Ukraine, a number of agri-food enterprises are moving to Industry 4.0. The most innovative companies in Ukraine are the largest exporters "Kernel", "MHP", "ASTARTA-KYIV". "Kernel", a large producer and exporter of sunflower oil, has been successfully implementing innovations for the agro-industrial complex of Ukraine for many years. The company uses digital technologies at all stages—from growing products to sales. The company's IT team digitized logistics, trading, and document management.

Citation: Tul, S.; Samoilyk, I.; Klymenko, V.; Shkurupii, O. Transformation of the Ukrainian Agri-Food Industry in the Context of Global Digitalization. *Eng. Proc.* **2023**, *40*, 26. https://doi.org/10.3390/engproc2023040026

Academic Editor: Hana Trollman

Published: 4 August 2023

All information about the processes taking place in agri-food production is collected in the "Kernel DigitalAgriBusiness" innovative ecosystem. "MHP", the largest producer and exporter of chicken in Ukraine, continues to use biogas to produce electricity, industrial steam, and heating. "MHP" biogas projects are a significant contribution to the company's energy independence and environmental responsibility. "ASTARTA-KYIV", a vertically integrated agricultural holding, developed a complex system of IT solutions for agribusiness management "AgriChain", which includes management of the land bank of the agricultural company (AgriChain Land), agricultural production (AgriChain Farm), monitoring of crops (AgriChain Scout), logistics of goods (AgriChain Logistics), warehouse management (AgriChain Barn), business processes (AgriChain Kit). Digital transformations are also being followed in the dairy industry. "Bel Shostka Ukraine" company is engaged in the digital transformation of the milk harvesting process. According to our research, breakthrough innovations are predominantly implemented by large Ukrainian agri-food companies, since they have significant financial resources for R&D, while SMEs are concentrating their efforts on the digitalization of business operations and implementation of energy efficient technologies.

Keywords: agri-food industry; digitalization; export; food security; innovative technologies; transformation

---

Author Contributions: Conceptualization, S.T.; methodology, I.S. and O.S.; formal analysis, S.T. and V.K.; investigation, S.T. and V.K.; resources, O.S.; writing—original draft preparation, S.T.; writing—review and editing, I.S. All authors have read and agreed to the published version of the manuscript.

Funding: This research received no external funding.

Institutional Review Board Statement: Not applicable.

Informed Consent Statement: Not applicable.

Data Availability Statement: Data available in a publicly accessible repository that does not issue DOIs. Publicly available datasets were analyzed in this study. This data can be found here: https://ukrstat.gov.ua/ (accessed on 1 June 2023).

Conflicts of Interest: The authors declare no conflict of interest.

engineering proceedings

MDPI

Abstract

Agriculture through Industry 4.0: Management, Challenges, and Opportunities in Hostile Environment: The Case of Iraq †

Sanar Muhyaddin

Faculty of Social and Life Sciences, Wrexham University, Mold Road, Wrexham LL11 2AW, UK; sanar.muhyaddin@glyndwr.ac.uk

† Presented at the International Conference on Industry 4.0 for Agri-food Supply Chains: Addressing Socio-economic and Environmental Challenges in Ukraine, Leicester, UK and Online, 24–25 July 2023.

Abstract: This paper reviews the future trends, the present situation and the prospects facing the application of Industry 4.0 technologies (e.g., Drones, texting technologies, GPS, etc.) in Iraq's agricultural sector in general and small agribusiness industries in particular. Iraq struggles with chronic structural and emerging challenges that have hindered its food production over the years. In 2019, Iraq's population had increased to around 39 million people, compared to 23.5 million Iraqis in 2000. This shift amounts to a 66% increase in population in 20 years. Food supply, whether locally produced or imported, has been struggling to catch up with the population growth, and Iraq has become increasingly dependent on food imports to meet domestic demand. Between 1985 and 2017, food imports increased from USD 2 billion to USD 11 billion, growing from 19 to 21% of total imports. Iraq's economy has been highly dependent on the oil sector; with declining oil prices, politicians and international communities have emphasised that modern and smart agriculture that applies Industry 4.0 technology can increase productivity and be a source of job creation, income generation and self-reliance. Nonetheless, evidence obtained via this exploratory study using unstructured interviews (3 out 33) via an interpretive approach and conducting thematic analysis of the interviews with farmers and agri-entrepreneurs indicate that there is a long journey ahead before Iraq can rely on agriculture and new technologies, instead of oil, for its economy and to improve its food system. Moreover, Iraq's political turmoil and uncertainty, the cyclical conflict and wars, and the corruption and mismanagement of state resources exacerbate this problem. Indeed, farmers face many challenges, such as a lack of infrastructure and security, the dominance of state-owned enterprises, and financing issues, with many farm owners struggling to access finance since there is no clear banking or credit system. On top of that issue, the diminishing of tacit knowledge among the farmers and the labour shortage issue in the agricultural sector hinders application of smart technology, as the majority of workers remaining in the market are unskilled labour. This study contributed to the literature on the application of smart technologies in agriculture and its socioeconomic effects. The exploratory nature of this study identifies areas for future research. Additionally, the arguments presented in this research highlight the challenges farmers and agri-entrepreneurs face in adopting new technologies and the tactics that they use to survive in a hostile environment like Iraq. It will also offer suggestions to help policymakers and international communities focus on intervention to help farmers and agri-entrepreneurs in Iraq to improve their performance.

Keywords: agricultural technologies; Industry 4.0; Industry 5.0; smart technology and agri-entrepreneurs; food system; Industry 4.0 in a war zone

Citation: Muhyaddin, S. Agriculture through Industry 4.0: Management, Challenges, and Opportunities in Hostile Environment: The Case of Iraq. *Eng. Proc.* **2023**, *40*, 27. https://doi.org/10.3390/engproc2023040027

Academic Editor: Hana Trollman

Published: 14 August 2023

Funding: This research received no external funding.

Institutional Review Board Statement: Not applicable.

Informed Consent Statement: Not applicable.

Data Availability Statement: Not applicable.

Conflicts of Interest: The author declares no conflict of interest.

engineering proceedings

MDPI

Abstract

Understanding the Geopolitical and Socio-Economic Factors Affecting the Food Supply Chain in Ukraine: An Exploratory Study †

Al-Amin Abba Dabo *, Olalekan Adisa, Rahul S Mor and Dilshad Sarwar

Business Systems & Operations, Faculty of Business and Law, University of Northampton, Northampton NN1 1SP, UK; olalekan.adisa@northampton.ac.uk (O.A.); rahul.mor@northampton.ac.uk (R.S.M.); dilshad.sarwar@northampton.ac.uk (D.S.)

* Correspondence: al-amin.dabo@northampton.ac.uk

† Presented at the International Conference on Industry 4.0 for Agri-food Supply Chains: Addressing Socio-economic and Environmental Challenges in Ukraine, Leicester, UK and Online, 24–25 July 2023.

Abstract: This paper presents an ongoing study that explores the influence of geopolitical and socio-economic factors on the Ukrainian food supply chain and its implications for food security. Focusing on specific sectors within Ukraine, our research aims to provide a comprehensive understanding of the intricate dynamics between geopolitics, socio-economic conditions, and the functioning of the food supply chain. To collect data, we have employed focus groups as the primary methodology, engaging key stakeholders from various sectors, including farmers, distributors, retailers, policymakers, and consumers. These focus group discussions enable us to delve into their perspectives, experiences, and challenges in relation to the influence of geopolitical and socio-economic factors on the Ukrainian food supply chain. Preliminary literature review reveals several noteworthy insights, including the impact of trade policies and regional conflicts on the availability and accessibility of specific food products in targeted regions of Ukraine. Building upon these initial findings, our ongoing study aims to propose strategies to enhance the resilience and efficiency of Ukraine's food supply chain. By tailoring policies to address the specific needs of different regions and socio-economic groups, we anticipate mitigating the adverse effects of geopolitical dynamics on the food system. Moreover, fostering collaboration among stakeholders will be crucial in navigating the complexities and challenges inherent in managing the Ukrainian food supply chain.

Keywords: food supply chain; food security; Ukraine; geopolitical factors; socio-economic factors; resilience strategies

Citation: Dabo, A.-A.A.; Adisa, O.; Mor, R.S.; Sarwar, D. Understanding the Geopolitical and Socio-Economic Factors Affecting the Food Supply Chain in Ukraine: An Exploratory Study. *Eng. Proc.* **2023**, *40*, 28. https://doi.org/10.3390/engproc2023040028

Academic Editor: Hana Trollman

Published: 15 August 2023

Author Contributions: Conceptualization, R.S.M. and O.A.; methodology, A.-A.A.D.; validation, R.S.M., A.-A.A.D. and O.A.; investigation, A.-A.A.D.; resources, D.S.; writing—original draft preparation, A.-A.A.D.; writing—review and editing, D.S. and O.A. All authors have read and agreed to the published version of the manuscript.

Funding: This research received no external funding.

Institutional Review Board Statement: Not applicable.

Informed Consent Statement: Informed consent was obtained from all subjects involved in the study.

Data Availability Statement: This is an ongoing study and as such no new data has been created as of yet.

Conflicts of Interest: The authors declare no conflict of interest.

engineering proceedings

Proceeding Paper

Being Resilient in Challenging Times in Food Supply Chains †

Louise Manning

Lincoln Institute for Agri-Food Technology, University of Lincoln, Riseholme Park, Lincoln LN2 2BJ, UK; lmanning@lincoln.ac.uk

† Presented at the International Conference on Industry 4.0 for Agri-food Supply Chains: Addressing Socio-economic and Environmental Challenges in Ukraine, Leicester, UK and Online, 24–25 July 2023.

Abstract: Resilience has been described in many ways and in this paper is considered as the ability of an organization or wider food supply chain to not just 'bounce back' to a steady state, but to 'bounce forward' or 'bounce without breaking' to a new sense of becoming that is continually reshaped by the evolving environment. This keynote-positioned conceptual paper reflects aspects of resilience that will be considered during the conference. Resilience as a concept can be framed in the context of food supply chains in terms of the triad of natural systems resilience, organizational resilience, and personal resilience. The main focus of this paper is on organizational resilience and developing the capacity to buffer shocks and to adapt to changing supply chain drivers and demands as the world in which the organization operates becomes more uncertain. One infrastructural approach to improving adaptive capacity, as a result resilience, is the integration of digital technologies and smart systems into food supply chains, what is often described as Industry 4.0.

Keywords: resilient; food; supply chains; food system; shocks; challenges; complexity; buffer; adaptiveness; Industry 4.0

Citation: Manning, L. Being Resilient in Challenging Times in Food Supply Chains. *Eng. Proc.* **2023**, *40*, 1. https://doi.org/10.3390/engproc2023040001

Academic Editor: Hana Trollman

Published: 14 July 2023

1. Introduction

Resilience as a concept has been defined in many ways in the literature depending firstly on the initial academic discipline from which the application arises (engineering, ecology, business management and so forth) and how the conceptualization is then applied to food supply chains in particular. First, there is the "business as usual" view of resilience, i.e., after a supply chain shock, resilience is the ability of a business, supply chain or a system to return to its original stable state [1–4]. Alternatively, the "bounce forward" view of an organization focuses on resilience as being the taking of opportunity after a disturbance and to transition to a new, more desirable state [1], and then resilience is conceptualized as an organization being perceived as being resilient when it has the transforming capacity to change, renew and grow [5–7].

Generally, considerations of resilience reflect an understanding that there is no steady state, that systems, especially food systems, are dynamic and ever changing, and operate within an environment that is uncertain [6], and increasingly so. Therefore, embedding the capacity to adapt and transform to new often non-linear conditions is essential [5,6]. Indeed, organizations, or whole supply chains, that have the core competencies to capture opportunities that arise from system disturbance and outperform others who fail to adapt can be described as being resilient [8,9] rather than being vulnerable [10], i.e., to 'bounce without breaking' [9].

Organizations, food supply chains and wider food systems are intrinsically complex and diverse. Food systems are composed of multiple activities, processes, value chains, actors, interactions both internal and external to the commercial relationships, and the impact of these activities affect multiple stakeholders in diverse and sometimes conflicting ways [11]. Therefore, food system resilience can be described as "the capacity over time of a food system, and its units at multiple levels, to provide sufficient, appropriate and accessible food to all, in the face of various and even unforeseen disturbances" [11] (p. 19). This definition suggests both a spatial aspect and a temporal aspect to resilience, in that shocks can be singular or multiple, both over time, and in their single or multiple foci in a food system. Shocks, economic disruption, wars and conflicts, harvest failure, supply issues, consumer boycotts of particular foods or organizations, and so forth, can influence, at different levels, the food system. Macro shocks define those shocks that influence at the supranational, national, regional or supply chain level; meso shocks at the organizational level; and micro shocks at the individual and personal level [12]. The multiplicity and complexity of interactions of such shocks is beyond exploration in this short paper.

Multiple shocks, or their associated ascribed magnitude of risk, can come together in a non-linear, complex event to produce an accumulated magnitude of risk and associated impact(s) that collectively could be greater than if the individual risks had occurred independently [13]. This non-linear aggregated and complex event has been termed a "perfect storm" [14]. Ref. [13] (p. 291) describes a perfect storm as "a combination of uncertainty, and aggregated "risky events with singular and multiple negative outcomes occurring simultaneously." Therefore, whilst risk assessment processes can assess single, independent risks associated with shocks and prioritize them in terms of their management and possible elimination, the level of dependency and interdependency between these events is difficult to determine [14], and, as a result, the aggregated risk of a multiple shock event occurring, its severity should it occur and the impact on the organization and wider supply chain is difficult to accurately compute, if such models actually exist.

Resilience, buffer capacity, adaptive capacity and redundancy are important elements. Buffer capacity represents the ability of an organization to utilize its resources (financial, physical, natural, human and social) to withstand a shock [8,9]. Adaptive capacity is the ability to adapt and change in the event of a shock, and the third element redundancy or substitutability [11] is also important. There are two elements to redundancy: firstly, being able to anticipate a shock/simultaneous shocks; and secondly, being prepared for a disruptive event, e.g., by having a business continuity plan in place [15]. The main focus of this paper is on organizational resilience and developing the capacity to buffer shocks and to adapt to changing supply chain drivers and demands. Section 2 explores adaptive capacity, panarchy and resilience in more detail and Section 3 provides some concluding thoughts.

2. Adaptive Capacity

Adaption can occur at the enterprise, organizational or supply chain level and involves a realigning, reconfiguring or reassorting of resources requiring the sacrificing of individual aspects of practice to deliver overall organizational resilience [6]. These adaption responses can be planned, or can be spontaneous, and can either be short-term tactics or long-term strategies. Thus, there is a difference between potentially transient adaptive responses and transformative activities that lead to long-term resilience [5,6]. The concept of panarchy positions resilience in terms of multiple interventions and associated interactions (both intended and unintended; positive and negative) that operate at a range of inter-level and intra-level loci (micro, meso, macro), horizontally and vertically (top-down, bottom-up, or peer-to-peer, often simultaneously), and where thresholds exist that can trigger feedback loops [16,17].

A panarchy forms as an interaction of adaptive cycles interlinked and crosslinked "across different levels on scales of time, space, and meaning," at the supply chain level, the political-economic level, and the planetary level [18] (p. 58). Two main cross-linkages are described by [18] as 'remember' and 'revolt' that interface with elements of panarchy, including reorganisation, exploitation, conservation and release [19–21]. Thus, the adaptive cycle is formed of four phases: growth (exploitation), conservation, collapse (release) and reorganization [22]. The concept of panarchy reflects system adaptability and system connectedness, but a further element of resilience is system-level temporal and spatial trade-offs, e.g., temporal aspects of using resources now or conserving for the future and spatial aspects such as distance between actors in a food supply chain and the impact of resilience on the situational characteristics of suppliers and the logistical implications [23]. These examples show the need for a more expansive and adaptive framing of food supply chains rather than a reductionist and static view [18]. Organizational and supply chain resilience relies on socio-technical mechanisms of release and reorganization, learning, relearning, unlearning, [24] acceptance or revolt, especially with the uncertainty created by contemporary shocks such as the COVID-19 pandemic, the Ukraine–Russia conflict, climate change, global warming, the rate of technological and institutional change, innovation or even revolution [25], both implicitly, and also how mitigation processes themselves impact food supply chains. In essence, for food supply chains to be resilient, they need to be less static, ridged and brittle, and instead more responsive, adaptive and able to buffer and adapt to shocks. Connectedness can both improve resilience, but be a vulnerability in of itself, if connectedness drives rigidity [25]; so, when considering 'bouncing back' or 'bouncing forward', the requisite state of equilibrium and what that 'looks like' is important [26].

Diversity supports ecological and socio-economic framings of resilience [23]. Diversity is the degree of variability in terms of functional diversity (integral components or functions [23]), and cognitively the variability through strategic responses to positive and negative triggers, internal or external to the system. Variety and diversity influence organizational equilibria, the thresholds at which shock(s) impact(s), and the application of remembering as a response mechanism. Constraints include organizational memory influencing adaption [26], the variety of outcomes and how they are influenced by the variety of activities employed to deliver these outcomes [27]. The introduction of Industry 4.0 into food supply chains has been positioned as a means to mitigate supply chain disruption through increasing agility [28]. Being adaptive and being resilient are a complex interaction of these interconnected mechanisms that are heavily influenced not only by the components and their functionality within the system, but also how those components are connected within socio-economic and socio-technical levels, and how they are connected across these levels.

3. Concluding Thoughts

Resilience has been described in many ways and in this paper is considered as the ability of an organization or wider food supply chain to not just 'bounce back' to a steady state, but to 'bounce forward' to a new status that is shaped by the evolving environment. Organizational resilience at the system level is presented in a non-linear and non-static view with the system being seen as "becoming" in terms of transformational aspects rather than simply "being" with regard to its transactional aspects [18]. Thus, resilience as a state and in terms of organizational thinking requires organizations, and indeed whole supply chains, to develop mechanisms that are firstly framed by the structures and connections, both physical and digital, that allow the organizations to operate in a dynamic environment, secondly, utilizes organizational memory and knows when to disregard it when it is a barrier to survival, and, finally, can embrace and grow through revolution whether it be social, technological or geopolitical. This notion of resilience reflects that there is no steady state, no business as usual, that 'bouncing back' or 'bouncing without breaking' is not a position which embraces all opportunity, and even 'bouncing forward' is not an end in itself, but only a transient position.

Funding: This research received no external funding.

Institutional Review Board Statement: Not applicable.

Informed Consent Statement: Not applicable.

Data Availability Statement: Not applicable.

Conflicts of Interest: The author declares no conflict of interest.

References

1. Christopher, M.; Peck, H. Building the resilient supply chain. *Int. J. Logist. Manag.* **2004**, *15*, 1–14. [CrossRef]
2. Morecroft, M.D.; Crick, H.Q.P.; Duffield, S.J.; Macgregor, N.A. Resilience to climate change: Translating principles into practice. *J. Appl. Ecol.* **2012**, *39*, 547–551.
3. Holling, C.S. Engineering resilience versus ecological resilience. In *Engineering within Ecological Constraints*; Schulze, P., Ed.; National Academy Press: Washington, DC, USA, 1996; pp. 31–44.
4. Pimm, S.L. *The Balance of Nature? Ecological Issues in the Conservation of Species and Communities*; University of Chicago Press: Chicago, IL, USA, 1991.
5. Walker, B.; Holling, C.S.; Carpenter, S.R.; Kinzig, A. Resilience, adaptability and transformability in social–ecological systems. *Ecol. Soc.* **2004**, *9*, 5.
6. Darnhofer, I. Strategies of family farms to strengthen their resilience. *Environ. Policy Gov.* **2010**, *20*, 212–222.
7. Manning, L.; Soon, J.M. Development of sustainability indicator scoring (SIS) for the food supply chain. *Br. Food J.* **2016**, *118*, 2097–2125.
8. Shadbolt, N.M.; Olubode-Awosola, F. Resilience, Risk and Entrepreneurship. *Int. Food Agribus. Manag. Rev.* **2016**, *19*, 33–52. Available online: http://www.ifama.org/files/IFAMR/Vol19/Iss2/220140117.pdf (accessed on 11 June 2023).
9. Shadbolt, N.; Olubode-Awosola, F.; Rutsito, B. Resilience to 'bounce without breaking' in New Zealand dairy farm businesses. *IFMA 19 Transform. Agric.* **2013**, *2*, 1–14. Available online: http://ifmaonline.org/wp-content/uploads/2014/08/13_Shadbolt_etal_P92-105v2.pdf (accessed on 11 June 2023).
10. Folke, C. Resilience: The emergence of a perspective for social-ecological systems analyses. *Glob. Environ. Change* **2006**, *16*, 253–267.
11. Tendall, D.M.; Joerin, J.; Kopainsky, B.; Edwards, P.; Shreck, A.; Le, Q.B.; Kruetli, P.; Grant, M.; Six, J. Food system resilience: Defining the concept. *Glob. Food Sec.* **2015**, *6*, 17–23.
12. Higashi, S.Y.; de Queiroz Caleman, S.M.; de Aguiar, L.K.; Manning, L. What causes organizations to fail? A review of literature to inform future food sector (management) research. *Trends Food Sci. Technol.* **2020**, *101*, 223–233. [CrossRef]
13. Manning, L.; Birchmore, I.; Morris, W. Swans and Elephants: A typology to capture the challenges of food supply chain risk assessment. *Trends Food Sci. Technol.* **2020**, *106*, 288–297. [CrossRef] [PubMed]
14. Paté-Cornell, E. On "black swans" and "perfect storms": Risk analysis and management when statistics are not enough. *Risk Anal. Int. J.* **2012**, *32*, 1823–1833. [CrossRef] [PubMed]
15. Grûzauskas, V.; Vilkas, M. Managing capabilities for supply chain resilience through it integration. *Econ Bus.* **2017**, *31*, 30–43. [CrossRef]
16. Gunderson, L.H.; Holling, C.S. (Eds.) *Panarchy. Understanding Transformations in Human and Natural Systems*; Island Press: Washington, DC, USA, 2002.
17. Berkes, F.; Ross, H. Panarchy and community resilience: Sustainability science and policy implications. *Environ. Sci. Pol.* **2016**, *61*, 185–193.
18. Wieland, A. Dancing the supply chain: Toward transformative supply chain management. *J. Supply Chain. Manag.* **2021**, *57*, 58–73. [CrossRef]
19. Allen, C.R.; Angeler, D.G.; Garmestani, A.S.; Gunderson, L.H.; Holling, C.S. Panarchy: Theory and application. *Ecosys* **2014**, *17*, 578–589. [CrossRef]
20. Holling, C. Resilience of ecosystems: Local surprise and global change. In *Global Change: The Proceedings of A Symposium Sponsored by the Internat. Council of Scientific Unions (ICSU) during Its 20th General Assembly in Ottawa, Canada on September 25, 1984*; Malone, T.F., Ed.; Cambridge University Press: Cambridge, UK, 1985; pp. 228–269.
21. Häring, I.; Ganter, S.; Finger, J.; Srivastava, K.; Agrafioti, E.; Fuggini, C.; Bolleta, F. Panarchy process for risk control and resilience quantification and improvement. In Proceedings of the 30th European Safety and Reliability Conference and the 15th Probabilistic Safety Assessment and Management Conference (ESREL 2020–PSAM 15), Venice, Italy, 1–5 November 2020; Baraldi, P., Di Maio, F., Zio, E., Eds.; Research Publishing: Singapore, 2020.
22. Van Apeldoorn, D.F.; Kok, K.; Sonneveld, M.P.; Veldkamp, T. Panarchy rules: Rethinking resilience of agroecosystems, evidence from Dutch dairy-farming. *Ecol. Soc.* **2011**, *16*, 39. Available online: http://www.ecologyandsociety.org/vol16/iss1/art39/ (accessed on 10 June 2023). [CrossRef]
23. Kharrazi, A.; Fath, B.D.; Katzmair, H. Advancing empirical approaches to the concept of resilience: A critical examination of panarchy, ecological information, and statistical evidence. *Sustainability* **2016**, *8*, 935. [CrossRef]
24. Manning, L.; Morris, W.; Birchmore, I. Organizational unlearning: A risky food safety strategy? *Compr. Rev. Food Sci. Food Saf.* **2023**, *22*, 1463–2487. [CrossRef]
25. Gotts, N.M. Resilience, panarchy, and world-systems analysis. *Ecol Soc.* **2007**, *12*, 24. Available online: http://www.ecologyandsociety.org/vol12/iss1/art24/ (accessed on 3 June 2023). [CrossRef]
26. Ashby, W.R. Principles of the self-organizing system. In *Principles of Self-Organization: Transactions of the University of Illinois Symposium*; Von Foerster, H., Zopf, G.W., Jr., Eds.; Pergamon Press: London, UK, 1962; pp. 255–278.

27. Ashby, W.R. Variety, constraint, and the law of requisite variety. In *Modern Systems Research for the Behavioral Scientist*; Buckley, W., Ed.; Aldine Publishing Co.: Chicago, IL, USA, 1968.
28. Spieske, A.; Birkel, H. Improving supply chain resilience through industry 4.0: A systematic literature review under the impressions of the COVID-19 pandemic. *Comput. Ind. Eng.* **2021**, *158*, 107452. [PubMed]

engineering proceedings

MDPI

Proceeding Paper

Application of Modern Enterprise Resource Planning Systems for Agri-Food Supply Chains as a Strategy for Reaching the Level of Industry 4.0 for Non-Manufacturing Organizations †

Olena Kopishynska [1,*], Yurii Utkin [1], Ihor Sliusar [1], Volodymyr Muravlov [2], Khanlar Makhmudov [1] and Liudmyla Chip [3]

1 Educational and Scientific Institute of Economics, Management, Law and Information Technologies, Poltava State Agrarian University, 36039 Poltava, Ukraine
2 Faculty of Engineering and Technology, Poltava State Agrarian University, 36039 Poltava, Ukraine
3 Faculty of Accounting and Finance, Poltava State Agrarian University, 36039 Poltava, Ukraine
* Correspondence: olena.kopishynska@pdaa.edu.ua; Tel.: +38-095-809-34-50

† Presented at the International Conference on Industry 4.0 for Agri-food Supply Chains: Addressing: Socio-economic and Environmental Challenges in Ukraine, Leicester, UK, 24–25 July 2023.

Abstract: The Ukrainian Industry 4.0 strategy envisages the positioning of Ukraine as a high-tech, post-industrial country, integrated into global technological chains of value creation and producing unique engineering services and high-quality products. In particular, for the needs of the territorial communities of Ukraine in the conditions of war, the uninterrupted operation of agro-food supply chains and ensuring the ecological safety of these territories has become especially important. This paper explores the possibilities of creating a unified digital information space in a modern cloud-based enterprise resource planning (ERP) system to improve the management of all subjects in the territorial community and facilitate the transition to the Industry 4.0 technology landscape.

Keywords: Industry 4.0; Community 4.0; cloud ERP; agri-food; Internet of Things

Citation: Kopishynska, O.; Utkin, Y.; Sliusar, I.; Muravlov, V.; Makhmudov, K.; Chip, L. Application of Modern Enterprise Resource Planning Systems for Agri-Food Supply Chains as a Strategy for Reaching the Level of Industry 4.0 for Non-Manufacturing Organizations. *Eng. Proc.* **2023**, *40*, 15. https://doi.org/10.3390/engproc2023040015

Academic Editor: Hana Trollman

Published: 18 July 2023

1. Introduction

The 21st century has witnessed the emergence of Industry 4.0, which is characterized by the dominance of knowledge and digital data. At the beginning of the Industry 4.0 era, the focus was solely on industrial production, but the paradigm has since expanded to include other areas of human activity such as agriculture (Agriculture 4.0), healthcare (Healthcare 4.0), logistics (Logistics 4.0), energy (Energy 4.0), and elements of new educational technologies (Education 4.0) [1].

The creation of 1436 territorial communities in Ukraine is the result of a key pro-European reform of local governments. The Community 4.0 program is related to this. The goal is the implementation of digital projects to ensure sustainability, attractiveness for investments, and new concepts for non-industrial organizations during the war and recovery in the future [2]. The gradual creation of Industry 4.0 centered at universities, such as Poltava State Agrarian University (PSAU), allows enterprises, scientific institutions, state authorities, and public formations to be involved in the implementation and approval of projects.

Industry 4.0 is a continuation and expansion of Industry 3.0 through the framework of new technologies (artificial intelligence (AI), industrial Internet of Things (IIoT), analytics and processing, and others). The broader application of modern ERP class systems is considered a necessary element of the strategy to achieve the level of technology and Culture 4.0 for non-manufacturing companies and organizations [3]. The role of powerful ERP systems in processing new types of data, globalizing production, and decentralizing management remains debatable. ERP systems will have to solve new tasks, such as data correlation and managing larger and more complex volumes of data. This study employs data gathered during advisory consultations provided by scientists at PSAU [4,5]. These consultations were conducted in 18 different communities, and four pilot projects were discussed to establish a unified digital information space through modern ERP systems.

2. Methods

The authors applied methods of quantitative and qualitative analysis of software, business processes in research subjects. Collection of primary data is carried out on the basis of a questionnaire by interviewing leading specialists from different departments and organizations through agreed questionnaires in different territorial communities with the agreement of their leaders. A case study of a pilot project for the implementation of a unified resource management system by the executive bodies of territorial communities in Ukraine has been prepared. Processing and use of data were carried out during the discussion of pilot projects in a tripartite format: customer (community)—consultants (scientists from PSAU)—performers (IT companies).

3. Results

The executive committee of a territorial community manages several departments, some of which are similar to enterprises (such as accounting, finance, land, legal, and communal management). As a rule, a large number of agricultural enterprises (agroholdings, small- and medium-sized enterprises, and farms) operate within each territorial community, which are often the main sources of the territorial community's budget. In addition, the executive committees of territorial communities are entrusted with the responsibility of forming a food reserve for the relevant period.

This research showed that different departments and institutions use an uncoordinated set of software: accounting systems of various types and complexity, transport management systems, agricultural production management systems, medical systems, banking clients, and others. Therefore, it is necessary to form a single software ecosystem on an ERP platform that is more flexible to support development [6,7]. To customize the ERP system modules for specific organizations, IT companies create company database models based on Oracle technologies.

Hence, the latest version of the cloud-based "Universal 9.0" ERP system has been implemented, which incorporates all modern technologies for processing large volumes of data, user interfaces, and architectural solutions. The advanced architecture of ERP "Universal 9", built according to the principle of a multi-tier system, makes it possible to scale the system to most enterprises in the future, including those that plan to use special sensors to collect operational data from hundreds of meters using a combination of artificial intelligence and Internet of Things (AI + IoT) technologies based on the algorithms described in [8].

This approach allows us not only to integrate an analog infrastructure into a modern digital ecosystem within a certain smart concept (Smart Cities, Agriculture 4.0, IIoT, etc.) [6], but also to provide real-time information about various aspects of the processes of production, storage, and transportation of products. This is due to the following main factors. Firstly, processing large amounts of data and applying machine learning (ML) algorithms allows one to identify patterns and predict possible problems in agricultural production. Secondly, sensors installed on farms or other production facilities can collect data on the soil conditions, chemical compositions of fertilizers, or the quality of animal feed for the purposes of product certification (including confirmation of an organic origin). As a result, this contributes to the efficiency, safety, and transparency of the agri-food supply chain.

In general, in addition to the IoT, it is advisable to use other smart technologies to control agri-food supply chains, including the following: automation of the processes of collection, sorting, and packaging of products to reduce the risk of human error. The introduction of blockchain to ensure transparency and the absence of data falsification in the agri-food chain, which allows one to track the path of the product from the farm to the final sale of goods to the consumer. This guarantees trust and quality assurance. Radio Frequency Identification (RFID) technology allows real-time tracking of the condition, location, and all movements of packaged products. When integrated with the tools discussed above, this makes it impossible to attempt to replace, delete, or change accounting information in the ERP system. This approach is quite relevant for the certification of ecological and organic products based on the analysis of information about the place and time of their collection (production), storage conditions and location, delivery to the end user, etc. In turn, comparing AI-based information on the product variety (weight parameters, fruit maturity, etc.) and market conditions will allow for the optimization of logistics processes and the efficiency of management decisions.

4. Conclusions

Since many enterprises and organizations are moving towards the introduction of Industry 4.0 technologies, it is necessary to critically analyze and reform their automation of production and management processes at level 3.0 at the beginning. In this work, an example of combining tasks of different parts of a non-industrial organization on a single platform of the ERP system is considered. For executive committees of territorial communities, such systems will allow for high-quality data analytics and significantly increase the effectivity of control over production, distribution, quality, logistics, and food at all stages of the supply chain.

Further research should focus on developing a stack of technologies that can be integrated with the cloud ERP system, leading to the creation of a modern software landscape that aligns with the Industry 4.0 framework.

Author Contributions: Conceptualization, O.K., L.C. and Y.U.; methodology, O.K., Y.U. and K.M.; software, Y.U., V.M., O.K. and I.S.; validation, O.K., Y.U., I.S., L.C. and V.M.; formal analysis, O.K., Y.U., L.C. and K.M.; investigation, O.K., Y.U., K.M., I.S. and L.C.; resources, O.K., Y.U., V.M.; data curation, O.K., Y.U.; writing—original draft preparation, O.K., Y.U., V.M.; writing—review and editing, O.K. and Y.U.; supervision, O.K. and Y.U.; project administration, O.K. All authors have read and agreed to the published version of the manuscript.

Funding: This research received no external funding.

Institutional Review Board Statement: Not applicable.

Informed Consent Statement: Not applicable.

Data Availability Statement: Data on the results of administrative reform, the number of communities in Ukraine, characteristics of land funds were taken from the official government website https://decentralization.gov.ua/en/newgromada (accessed on 15 June 2023). Data on the technical characteristics of the described ERP system "Universal 9" is provided by the software developer IT company Soft.Pro (Ukraine) and is publicly available on the official website https://softpro.ua/en/home (accessed on 15 June 2023). The work contains references to the previous studies of the authors, which are the basis for continuation in this research. The composition of the frameworks of Industry 4.0 technologies and their characteristics are publicly available, in particular in scientific publications.

Conflicts of Interest: The authors declare no conflict of interest.

References

1. Bongomin, O.; Yemane, A.; Kembabazi, B.; Malanda, C.; Chikonkolo Mwape, M.; Sheron Mpofu, N.; Tigalana, D. Industry 4.0 Disruption and Its Neologisms in Major Industrial Sectors: A State of the Art. *J. Eng.* **2020**, *2020*, 1–45. [CrossRef]
2. Communities. Available online: https://decentralization.gov.ua/en/newgromada (accessed on 15 June 2023).
3. Queiroz, M.M.; Pereira, S.C.F.; Telles, R.; Machado, M.C. Industry 4.0 and digital supply chain capabilities: A framework for understanding digitalization challenges and opportunities. *Benchmarking* **2021**, *28*, 1761–1782. [CrossRef]
4. Kopishynska, O.P.; Utkin, Y.V.; Voloshko, S.V.; Sliusar, I.I.; Kartashova, O.G. Algorithm of creating of an efficient cooperation between universities, business companies and agriculture enterprises during studying and implementation of information systems. In Proceedings of the 2018 IEEE 9th International Conference on Dependable Systems, Services and Technologies (DESSERT), Kyiv, Ukraine, 24–27 May 2018; pp. 682–686. [CrossRef]

5. Kopishynska, O.; Utkin, Y.; Sliusar, I.; Slyusar, V.; Protas, N.; Barabolia, O. Professional-oriented training of specialists under implementation of cloud computing information systems in cooperation between universities and IT companies. In Proceedings of the 14th International Multi-Conference on Society, Cybernetics and Informatics (IMSCI 2020), Online, 13–16 September 2020; pp. 17–22. Available online: https://www.iiis.org/CDs2020/CD2020Summer/papers/EA790UO.pdf (accessed on 15 June 2023).
6. Kopishynska, O.; Utkin, Y.; Kalashnik, O.; Mororz, S.; Lyashenko, V.; Barabolia, O.; Kartashova, O. Information Systems and Technologies in Agronomy and Business: Employers' Requirements-Oriented Study in Agricultural Universities. In Proceedings of the 25th World Multi-Conference on Systemics, Cybernetics and Informatics (WMSCI 2021), Virtual, 18–21 July 2021; pp. 113–118.
7. Kopishynska, O.; Utkin, Y.; Galych, O.; Marenych, M.; Sliusar, I. Main Aspects of the Creation of Managing Information System at the Implementation of Precision Farming. In Proceedings of the 2020 IEEE 11th International Conference on Dependable Systems, Services and Technologies (DESSERT), Kyiv, Ukraine, 14–18 May 2020; pp. 404–410. [CrossRef]
8. Slyusar, V.; Sliusar, I.; Bihun, N.; Piliuhin, V. Segmentation of analogue meter readings using neural networks. In Proceedings of the 11th International Workshop on Modern Machine Learning Technologies and Data Science (MOMLET&DS2022), Leiden, The Netherlands, 23–25 November 2022; pp. 165–175. Available online: https://www.slyusar.kiev.ua/Slyusar_MoMLeTDS_2022_6.pdf (accessed on 15 June 2023).

engineering proceedings

MDPI

Proceeding Paper

A Vision of the Food System, 2045 CE: Materiality Methods Can Define What Is Resilient and Critical †

Wayne Martindale [1,*] and Tom Æ Hollands [2]

[1] Food Insights and Sustainability National Centre for Food Manufacturing, University of Lincoln, 2 Peppermint Way, Holbeach PE12 7FJ, UK
[2] Food Science and Innovation Department Raynor Foods Ltd., Farrow Road, Widford Industrial Estate, Chelmsford, Essex CM1 3TH, UK; tom.hollands@sandwiches.uk.net
* Correspondence: wmartindale@lincoln.ac.uk
† Presented at the International Conference on Industry 4.0 for Agri-Food Supply Chains: Addressing Socio-Economic and Environmental Challenges in Ukraine, Leicester, UK and Online, 24–25 July 2023.

Abstract: This keynote and future paper summarises the methods that are being utilised to define the requirements of national populations' natural resource demands for the consumption of food and beverage products so that strategies may be more effectively developed to deal with crises. The methods are presented as part of a material analysis of the UK food supply system, and they are used to demonstrate sustainable practices in food manufacturing. Our current conclusions show that the limiting factors in the food system must be focussed on human-centred activities that interact with material flows, which are often overlooked in sustainability assessments. This is critical if we intend to tackle issues of security, resilience, and sustainability incisively.

Keywords: materiality; security; sustainability

Citation: Martindale, W.; Hollands, T.Æ. A Vision of the Food System, 2045 CE: Materiality Methods Can Define What Is Resilient and Critical. *Eng. Proc.* **2023**, *40*, 23. https://doi.org/10.3390/engproc2023040023

Academic Editor: Hana Trollman

Published: 1 August 2023

1. Introduction

This abstract is written for our keynote presentation at the International Conference on Industry 4.0 for Agri-food Supply Chains: Addressing Socio-economic and Environmental Challenges in Ukraine, held from 24 to 25 July 2023 at the University of Leicester, UK. This keynote further develops a previously published study [1] in which a narrative of food security was established in response to the current Russian State war on Ukraine.

Sustainable practices in the food system have been transformed by thought leaders associated with the delivery of the Green Revolution, Globalisation and Sustainable Development programmes. Each of these globally transformative programmes has fallen short of achieving their intended goals, and we report that this is because the measurement of progress has not been reactive enough in the face of crises. This reactivity has become more critical, and it has been emphasised in manufacturing practices because human values are not fully built into the applications of technology and extremely variable geopolitical conditions that continue to confound sustainable outcomes. Solutions have been identified by considering the production of life-securing services and products in the most hostile environments in which re-thinking the current manufacturing systems is a necessity [2]. The development of analytical methods that can provide timely assessments of materiality and criticality in food systems are required to deliver these technology solutions, and they have been previously tested using statistical data from the UN Food and Agricultural Organisation (FAO) [3]. These methods identify the most important food materials used by specific nations and thus develop a more focussed materiality of food systems and materials. The approach is similar to the approach used by engineers to assess materiality issues in product and market development, which consider the criticality of the supply of materials by mass, availability and environmental impact [4]. Digital technologies are known to be transformational in implementing these methods because they provide the means to develop increased reactivity and resilience. As such, they can transform the supply of agri-products, food ingredients and infrastructural materials used by the food system [5].

There are similarities in this approach to a Life Cycle Impact Assessment (LCIA), and the methods reported herein simplify the in-depth analysis used in LCIA and LCA so that reactive resilience can be deployed with confidence. The assessment of control, limits and materiality across the food system has had limited application. The specific methods for assessing the effectiveness of material flows are extremely well developed, but they have not been integrated for application. Notably, factories are where we begin to rigidly control the inefficiencies within energy or material flows, and an example of this is the development of new manufacturing strategies such as the production of fresh produce in Controlled Environment Agriculture (CEA) facilities—the so-called vertical farms. These are reforming manufacturing operations by reducing the criticality of specific materials, such as in the timely and projectable supply of fresh ingredients [2].

2. Analysis of Food System Data to Assess Supply Resilience

The assessment methodology tested and developed herein focusses on (1) commercial responses to supply chains impacted by climate change; (2) the utilisation of materials, including losses and waste; and (3) the supply of healthy diets. These three areas provide the interpretive focus for the assessment of food category data obtained from national food supply and balance databases so that a ranked assessment of food security is possible. Figure 1 shows how this method is initiated by assessing the critical supply of food categories at a national scale; this is achieved by calculating the ratio between the production quantity of the agri-food category and the Domestic Supply Quantity (DSQ) of the same category. The DSQ is calculated as the total national production mass and import mass, subtracting the total export mass. This means that a production-to-DSQ ratio value of one identifies that production is equal to supply, a ratio above one indicates that production exceeds supply and a ratio below one indicates that production is a limiting component of supply. This approach supports other database methodologies that identify limits to nutritional supply and can be used to support the application of LCA and carbon footprinting methods. Figure 1 shows the Production to DSQ ratio in terms of a Red Amber Green (RAG) risk register. Once the ratio is calculated, the data must be ranked, and Figure 1 shows an example of how risk can be segmented by thresholds for risk assessment, demonstrated herein this via a simplistic RAG register.

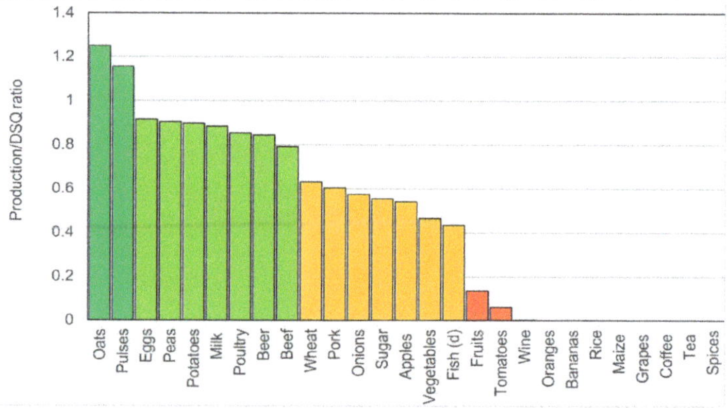

Figure 1. The ratio of UK Production to Domestic Supply Quantity (DSQ). DSQ is calculated as the addition of production and import mass, subtracting exports, to UK production in 2020. Data obtained from FAOStat Food Balance Databases. See Table S1 for the data utilized utilised to produce this figure.

3. Analysis of Food System Data to Assess Nutritional Resilience

Figure 2 develops these insights further by demonstrating how the ratio can be used to inform materiality in our food system, using the specific food category percentage supply of Calories and protein to the diet of UK citizens. This assessment is crucial because it provides an incisive view of valuable Calories and protein, which are part of the material value of the food system for specific food categories. The immediate outcome of the materiality shown in Figure 1, which demonstrates limitations to integrating the mass balance assessment shown in Figure 1 and the nutrition profiles shown in Figure 2 with carbon footprints and LCAs of the food categories. Figure 2 demonstrates this point well because the actual supplies of Calories and protein align well with what is considered a healthy and balanced diet; however, this is clearly not the case if national health outcomes are considered. As such, the impact of this analysis is the requirement to develop a greater diversity of food categories. Supplying Calories and protein in meals and diets must be a goal of UK food policy so that micronutrients and other factors are considered more effectively.

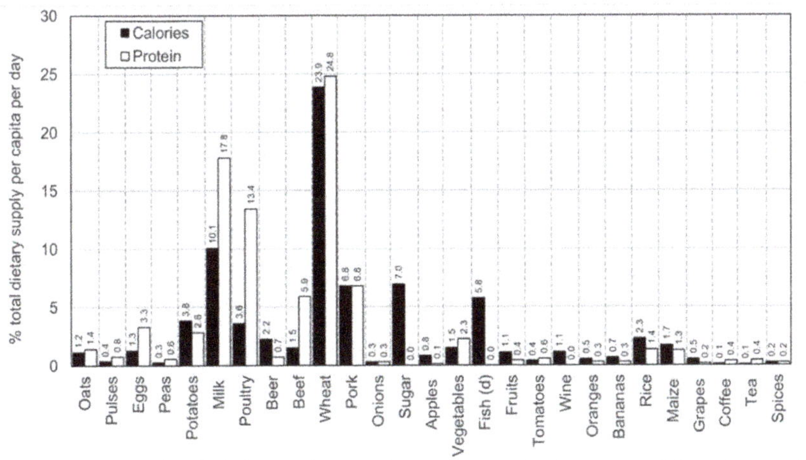

Figure 2. The percentage of total calorific and protein supply from different food categories in the UK in 2020, data obtained from FAOStat Food Balance Databases. See Table S1 for the data utilized utilised to produce this figure.

Figure 1 demonstrates the principle of first ranking those food categories that are most at risk, and this must be a pre-requisite for any security assessment that seeks to guide resilience in food supply. The consideration of materiality within the food system can help to enable sustainable practice in developing food supplies because food categories can be ranked by requirement, distribution effort and the services required to produce them [3]. Together they can define the criticality of specific food categories and the services that support their supply.

4. Analysis of Food System Data to Assess Logistical Resilience

Distribution and logistical operations are often critical limitations in securing sustainable food chains; an example of this is provided by what has happened in the Ukraine food system. Agricultural production has managed to continue during war and conflict since 2014 and most incredibly since 2022; this is likely to change dramatically in 2023. What has been crucial during conflict is that it is not only production capability but also the importance of transport from global regions; in 2023, there are critical signs that reductions in production will have a global impact. Table 1 shows that agricultural production alone is not enough to provide security, and critical distribution routes, such as shipping through the Black Sea ports of Ukraine, are critical to the distribution of agricultural commodities.

Table 1. The top ten ranked nations importing Ukrainian agricultural products from 31 July 2022 to 27 June 2023. The data were derived from the Humanitarian Data Exchange (2023) Black Sea Grain Initiative Vessel Movements, https://data.humdata.org/dataset/black-sea-grain-initiative-vessel-movements (accessed on 1 July 2023).

Departure Port	Country	Commodity	Tonnage	Departure
Chornomorsk	China	Maize	75,790	2 June 2023
Yuzhny/Pivdennyi	Spain	Wheat	74,904	18 January 2023
Yuzhny/Pivdennyi	The Netherlands	Maize	74,500	30 August 2022
Yuzhny/Pivdennyi	China	Maize	72,600	24 December 2022
Yuzhny/Pivdennyi	China	Maize	71,970	27 December 2022
Yuzhny/Pivdennyi	China	Maize	71,500	18 February 2023
Chornomorsk	China	Maize	71,500	15 January 2023
Yuzhny/Pivdennyi	China	Maize	71,500	24 December 2022
Yuzhny/Pivdennyi	Spain	Wheat	71,500	22 December 2022
Yuzhny/Pivdennyi	Indonesia	Wheat	71,400	14 December 2022

In geo-political conflict and war, the most vulnerable communities and nations are the most impacted, and Table 1 shows that the shipping of agricultural products from Black Sea ports in the past year for wheat and maize exceeded 70,000 tonnes per ship. Notably, Table 2 shows that similar tonnages will be transported to more Calorie-deficient regions of the world that are dependent on Ukrainian export, and these shipped cargoes all carry wheat. The data shown in Tables 1 and 2 are from the Humanitarian Data Exchange, and they do not provide a secure outlook for global food security within the next 1–5 years. Further data showing the impact of agri-transport agreements during the Russian war are available and demonstrate the critical requirement for logistical services to support production in order to deliver global food security [1].

Table 2. The first ten ranked low- and middle-income nations importing Ukrainian agricultural products from 31 July 2022 to 27 June 2023. The data were derived from the Humanitarian Data Exchange (2023) Black Sea Grain Initiative Vessel Movements, https://data.humdata.org/dataset/black-sea-grain-initiative-vessel-movements (accessed on 1 July 2023).

Departure Port	Country	Commodity	Tonnage	Departure
Chornomorsk	Sudan	Wheat	65,340	25 August 2022
Yuzhny/Pivdennyi	Yemen	Wheat	55,318	12 November 2022
Yuzhny/Pivdennyi	Yemen	Wheat	53,300	4 March 2023
Yuzhny/Pivdennyi	Yemen	Wheat	37,000	29 August 2022
Odesa	Sudan	Wheat	30,000	13 May 2023
Odesa	Ethiopia	Wheat	30,000	24 April 2023
Chornomorsk	Yemen	Wheat	30,000	24 March2023
Odesa	Ethiopia	Wheat	30,000	21 March 2023
Chornomorsk	Ethiopia	Wheat	30,000	21 January 2023
Chornomorsk	Ethiopia	Wheat	30,000	3 December 2022

5. For Discussion

We often cite the requirement for resilience in our food system, which is described as the ability to buffer and withstand rapid change. Conflict and war have prevented food security; policy narratives first made us aware of this in the 1970s, when they highlighted war and conflict in Eastern Africa. We now continue to face the impact of climate change, so expediently identifying critical materiality has become an important role for scientists to champion; without this, we will fail. The current war has maintained agricultural production, but remote sensing data indicate that 2023 will be a year of an drastically reduced production in Ukraine, and famine will follow in North Africa and the Middle/Near East if action is not taken now. The studies and methods reported herein must be communicated. They utilise what are largely open data systems, and it is our duty as food scientists to make policy makers aware of what the evidence is telling us. We can do this more openly and expediently than ever before, and it must be a common goal of our scientific community. The most important application of these methods and research is to optimise the use of energy in the food supply and bring climate change strategies into the industry that will prepare us for crises and enhance workforce understanding of these practices.

Supplementary Materials: The following supporting information can be downloaded at: https://www.mdpi.com/article/10.3390/engproc2023040023/s1, Table S1: The data utilized to produce Figures 1 and 2.

Author Contributions: Conceptualization, W.M. and T.Æ.H.; methodology, W.M.; software, W.M.; validation, W.M. and T.Æ.H.; formal analysis, W.M. and T.Æ.H.; investigation, W.M. and T.Æ.H.; resources, W.M. and T.Æ.H.; data curation, W.M. and T.Æ.H.; writing—original draft preparation, W.M.; writing—review and editing, W.M. and T.Æ.H.; visualization, W.M.; supervision, W.M.; project administration, W.M. and T.Æ.H. All authors have read and agreed to the published version of the manuscript.

Funding: This research received no external funding.

Institutional Review Board Statement: Not applicable.

Informed Consent Statement: Not applicable.

Conflicts of Interest: The authors declare no conflict of interest.

References

1. Jagtap, S.; Trollman, H.; Trollman, F.; Garcia-Garcia, G.; Parra-López, C.; Duong, L.; Martindale, W.; Munekata, P.E.S.; Lorenzo, J.M.; Hdaifeh, A. The Russia-Ukraine Conflict: Its Implications for the Global Food Supply Chains. *Foods* **2022**, *11*, 2098. [CrossRef] [PubMed]
2. Hollands, T.; Swainson, M.; Martindale, W. Food production in hostile environments. *Food Sci. Technol.* **2022**, *36*, 18–23.
3. Martindale, W.; Swainson, M.; Choudhary, S. The Impact of Resource and Nutritional Resilience on the Global Food Supply System. *Sustainability* **2020**, *12*, 751. [CrossRef]
4. Graedel, T.E.; Harper, E.M.; Nassar, N.T.; Nuss, P.; Reck, B.K. Criticality of metals and metalloids. *Proc. Natl. Acad. Sci. USA* **2015**, *112*, 4257–4262. [CrossRef] [PubMed]
5. Martindale, W.; Hollands, T.Æ.; Jagtap, S.; Hebishy, E.; Duong, L. Turn-key research in food processing ad manufacturing for reducing the impact of climate change. *Int. J. Food Sci. Technol.* **2023**. [CrossRef]

MDPI
St. Alban-Anlage 66
4052 Basel
Switzerland
www.mdpi.com

Engineering Proceedings Editorial Office
E-mail: engproc@mdpi.com
www.mdpi.com/journal/engproc

www.ingramcontent.com/pod-product-compliance
Lightning Source LLC
Chambersburg PA
CBHW062107220926
51850CB00010B/2923
* 9 7 8 3 0 3 6 5 9 3 8 8 3 *